tredition

tredition was established in 2006 by Sandra Latusseck and Soenke Schulz. Based in Hamburg, Germany, tredition offers publishing solutions to authors and publishing houses, combined with worldwide distribution of printed and digital book content. tredition is uniquely positioned to enable authors and publishing houses to create books on their own terms and without conventional manufacturing risks.

For more information please visit: www.tredition.com

TREDITION CLASSICS

This book is part of the TREDITION CLASSICS series. The creators of this series are united by passion for literature and driven by the intention of making all public domain books available in printed format again - worldwide. Most TREDITION CLASSICS titles have been out of print and off the bookstore shelves for decades. At tredition we believe that a great book never goes out of style and that its value is eternal. Several mostly non-profit literature projects provide content to tredition. To support their good work, tredition donates a portion of the proceeds from each sold copy. As a reader of a TREDITION CLASSICS book, you support our mission to save many of the amazing works of world literature from oblivion. See all available books at www.tredition.com.

 Project Gutenberg

The content for this book has been graciously provided by Project Gutenberg. Project Gutenberg is a non-profit organization founded by Michael Hart in 1971 at the University of Illinois. The mission of Project Gutenberg is simple: To encourage the creation and distribution of eBooks. Project Gutenberg is the first and largest collection of public domain eBooks.

Guano A Treatise of Practical Information for Farmers

Solon Robinson

Imprint

This book is part of TREDITION CLASSICS

Author: Solon Robinson
Cover design: Buchgut, Berlin – Germany

Publisher: tredition GmbH, Hamburg - Germany
ISBN: 978-3-8472-3391-6

www.tredition.com
www.tredition.de

Copyright:
The content of this book is sourced from the public domain.

The intention of the TREDITION CLASSICS series is to make world literature in the public domain available in printed format. Literary enthusiasts and organizations, such as Project Gutenberg, worldwide have scanned and digitally edited the original texts. tredition has subsequently formatted and redesigned the content into a modern reading layout. Therefore, we cannot guarantee the exact reproduction of the original format of a particular historic edition. Please also note that no modifications have been made to the spelling, therefore it may differ from the orthography used today.

WITH A

BRIEF SYNOPSIS OF ITS HISTORY, LOCALITY, QUANTITY, METHOD OF PROCURING,
PROSPECT OF CONTINUED SUPPLY, AND PRICE; ANALYSIS OF
ITS COMPOSITION, AND VALUE AS A FERTILIZER,
OVER ALL OTHER MANURES.

"If the experience of the last few years has taught us one thing more certainly than another, it is the unfailing excellence of Guano for every kind of crop which requires manure."

CONTENTS

INTRODUCTION

PERUVIAN GUANO – ITS USES AND BENEFITS.

EFFECTS PRODUCED BY THE USE OF GUANO IN VIRGINIA.

DR. FAIRFAX'S EXPERIMENTS WITH GUANO.

MR. NEWTON'S EXPERIMENTS.

GUANO vs. MANURE – EFFECTS UPON HEAVY LAND.

OTHER WITNESSES IN VIRGINIA IN FAVOR OF GUANO.

GUANO IN NORTH CAROLINA.

EXPERIMENTS IN MARYLAND.

EVIDENCE OF THE DURABLE EFFECTS OF GUANO.

THE FIVE FIELD SYSTEM AND GUANO.

ED. REYNOLDS ESQ., OF BALTIMORE, ON THE VALUE OF GUANO.

GUANO IN DELAWARE.

GUANO IN PENSYLVANIA.

GUANO IN NEW JERSEY.

GUANO ON LONG ISLAND.

GUANO IN MASSACHUSETTS.

EXPERIMENTS BY MR. TESCHEMACHER.

DIRECTIONS AS TO QUANTITY AND MANNER OF APPLYING GUANO TO VARIOUS CROPS AND SOILS.

PLASTER WITH GUANO.

WHAT IS GUANO? — ITS HISTORY AND LOCALITY. — AMOUNT AND VALUE.

PERUVIAN GUANO — ITS LOCATION — OWNERSHIP — QUANTITY — VALUE — HOW PROCURED.

DOES GUANO PAY?

APPENDIX.

SUCCESSFUL EXPERIMENTS WITH GUANO ON LONG ISLAND.

INTRODUCTION

The rapidly increasing use of guano, in the United States, and the growing conviction upon the public mind, that it is the cheapest and best purchasable manure in the world, together with the fact of a great want of information among American farmers, as to the best mode of applying it to the soil, has induced the agents of the Peruvian Government for the sale of guano in the United States, to employ the author of this pamphlet to collect and publish such information.

It is hoped the favorably and well known name of the author, as an agricultural writer and traveller, together with his extended opportunities of witnessing the application and effect of guano upon the various soils and climates of this country, will give this work such a character, as to induce every improving farmer, gardener, or horticulturist, in America to give it a careful perusal. The author believes it will be found to contain all and much more than its title imports, and be of great value to every person using or dealing in guano; as the analysis, not only of the pure article is given, but that of several specimens of adulterated samples, so as to enable the farmer to avoid being cheated by base counterfeits.

The author will be much obliged to any gentleman who will furnish him for publication in future editions of this work, or in the columns of The Agricultor, any details of experiments in the use of Peruvian guano, which will be useful to the farmers of this country, as it is his desire, as well as the guano agents, to give them useful facts; not only to increase the sale, but the fertility of the land, and wealth of the owners.

With assurances to my friends that I have no other interest in the increased consumption of guano, I am most sincerely and respectfully

Your old Friend,

Solon Robinson.

New York, October 1852.

A TREATISE ON GUANO.

PERUVIAN GUANO—ITS USES AND BENEFITS.

Of all manures procurable by the American Farmer, guano from the rainless islands of Peru, is perhaps not only the most concentrated—the most economical to the purchaser—but by its composition, as we will show by analysis, the best adapted to all the crops cultivated in this country requiring manure. For wheat, especially, it is the one thing needful. The mineral constituents of cultivated plants, as will also be shown by analysis, are chiefly lime, magnesia, potash, soda, chlorine, sulphuric and phosphoric acid; all of which will be found in Peruvian guano. Nitrogen, the most valuable constituent of stable or compost manures, exists in great abundance in guano, in the exact condition required by plants to promote rapid vegetation. The concentration of all these valuable properties in the small bulk of guano, renders it particularly valuable to farms situated in districts unprovided with facilities of cheap transportation. In some hilly regions, it would be utterly impossible to make any ordinary manure pay for transportation. With guano the case is very different—one wagon will carry enough with a single pair of horses to dress 12 or 16 acres; while of stable manure it would require as many or more loads to each acre to produce the same effect.

But this is not the greatest advantage in the use of this fertilizer; the first application puts the land in such condition, that judicious after cultivation renders it continuously fertile by its own action of productiveness and reproductiveness of wheat, clover and wheat, by turning in the clover of one year for the wheat of the next, and by returning the straw back to the ground where it grew, spread open the surface to shade the plants of clover and manure its roots, which in turn manure the corn or wheat.

As a source of profit alone, we should recommend the continuous application of Guano; knowing as we do, from our extensive means of observation, that no outlay of capital ever made by the farmer, is so [Pg 6] sure and certain to bring him back good returns for his money, as when he invests it in this invaluable fertilizer for his impoverished soil. In proof of this, we shall give the reader of this little work a number of experiments made by some of the most improving farmers in Virginia and other States.

EFFECTS PRODUCED BY THE USE OF GUANO IN VIRGINIA.

In no other part of the world, perhaps, can the beneficial effects of Guano be more plainly seen than in the tide-water region of Virginia. In the counties of King George, Westmoreland, Richmond, Northumberland, Lancaster, in the northern neck, as the peninsula between the Potomac and Rappahanock is termed; thousands of acres of land so poor and worthless a few years ago, it was barely rated as property, are now annually producing beautiful crops of wheat, corn and clover, solely by the application of Guano. In the meantime, the discovery of such an easy means of improving a worn out and barren soil, has increased the money value of land three or four hundred per cent. This is not all. Heretofore, the only part of this district considered worth cultivation was the bottom land bordering the rivers and creeks; the forest land yielding scanty crops for two or three years after being cleared, scarcely paying for the labor, while its value was rated at from $1 to $4 per acre, and unsaleable at that. Since the introduction of Guano, it is found these forest lands, which are of a sandy, loamy character, and much more pleasant than the bottom lands to till, can be cultivated with equal or greater profit than the stiff lands upon the bottoms. The writer has seen repeatedly in the counties mentioned, luxuriant fields of wheat, corn and clover, while directly alongside of such crops, the ground was almost as bare of vegetation as the sea-shore sands, too poor, as the common expression is there, to bear poverty grass. And what produced this change? Simply a dressing of 200 lbs. of Guano to the acre.

DR. FAIRFAX'S EXPERIMENTS WITH GUANO.

In April 1850 the writer was on the farm of Dr. Fairfax of King George county, who was one of the first, if not quite the first person in that part of the State who ever made use of this substance as a manure; and his wheat was then so large that a good sized dog was hidden from view in running through the field; while upon a neighboring piece of land of exactly the same quality, sowed at the same time, the ground scarcely looked green; in fact, it was remarked at the time by way of contrast to the one field hiding a dog, that the other would not hide a chicken—indeed, an egg might have been seen as far as though no wheat was [Pg 7] growing upon the ground. Both fields were just alike, both plowed and sowed alike, without manure, except 200 lbs of Peruvian guano upon one, and that sure to bring fifteen or twenty bushels to the acre, while the other would not exceed three bushels.

One of his first trials was with the African, of which he applied 400 lbs. to the acre upon 27 acres, which would not produce three bushels of wheat to the acre, in its natural condition, but with this application, notwithstanding it was 32 per cent. water, and, consequently, had lost much of it ammonia, he made an average of 12¾ bushels to the acre on the whole field. Upon another, he increased the usual average yield from 8 to 18 bushels, while, in his opinion, the permanent improvement of the land was of greater value than the increased yield of the first crop; for now clover will grow where none would grow before; another advantage arising from guano is, the wheat ripens so much earlier (15th of June) it escapes the rust, so apt to blight that which is late coming to maturity. He now sows wheat in the fore part of September, three pecks to the acre, after having previously plowed in 200 lbs. of Peruvian guano to the acre, and after the first harrowing sows the clover seed. The land is a yellow clay loam, uneven surface, very much worn; in fact, without the guano, and with all the manure that could be made upon the farm—for no straw no manure—not worth cultivating. Dr. F. had been using guano three years, at the date of our visit, and thought his prospect good for a thousand bushels of wheat upon the same ground, which, without guano would not produce one hundred and fifty.

MR. NEWTON'S EXPERIMENTS.

The Hon. Willoughby Newton, of Westmoreland County, was one of the earliest and most successful experimenters in the use of guano in Virginia. He owns large and productive farms on the Potomac, but on account of the forest land being more healthy for a residence, he bought a tract of it for that purpose; not having any design of ever putting it into cultivation. In fact, it was so poor he could not. The manure of the farm, if it had not been wanted there, was several miles distant—too far to haul; and so the land lay an uncultivated, unprofitable barren waste around his fine mansion; but it did not lay so very long after he discovered the renovating power of guano. It is now annually covered with broad fields of wheat, from which he has realized upwards of twenty bushels to the acre; and the most luxuriant growths of clover upon which he can pasture any amount of stock he pleases, where three years previous a goat would have found difficulty in sus [Pg 8] taining life. Mr. Newton's first experiment—what was then an experiment is now a certainty—was made with African guano. But we will give the account of his operations in his own straight-forward, easily understood, farmer-like language.

"In the effect of *guano*, especially the Peruvian, I have never been disappointed. I have used it now for four years, with entire satisfaction having each year been induced to enlarge my expenditure, until last year it reached eight hundred dollars, and for the crop of wheat this fall it exceeds one thousand. I have observed with astonishment its effect in numerous instance on the poor "forest lands" alluded to in a former part of this address. What the turnip and sheep husbandry have done for the light lands of Great Britain, the general use of guano promises to do for ours. Lands a few years ago deemed entirely incapable of producing wheat, now produce the most luxuriant crops. From 15 to 20 bushels for one sowed, is the ordinary product on our poorest lands, from the application of 200 lbs. of Peruvian guano. I may remark, it is not usual, in Eastern Virginia, to sow more than a bushel of wheat to the acre, and that I deem amply sufficient. Upon this subject I hope a few details may not be considered tedious or uninteresting. I applied last fall $350 worth of guano, partly Peruvian and partly Patagonian, on a poor farm "in the forest," which cost a few years ago four dollars an acre, and reaped 1089 bushels of beautiful wheat from 78 sowed. Forty-six bushels were sowed on fallow, (both guano and wheat put in with the cultivator, followed by a heavy harrow,) and yielded 790 bushels or over 17¼ for one. A considerable part of this was dressed with Patagonian guano, and was much inferior to the other portion. A lot on which 15 bushels was sowed, and dressed with Peruvian guano, was threshed separately, and yielded 301 bushels, or over 20 for one. The whole cost of the farm was $1520, and I have good reason to expect with a favorable season from the crop now sowed and dressed with guano, a bushel of wheat for every dollar of the prime cost of the farm. Many other instances of profit from the use of guano, equally striking have occurred among my neighbors and friends, but I confine myself to those stated, because having come under my immediate observation, I can vouch for their entire accuracy. It has been frequently objected to the use of guano, that it is not permanent. It would be unreasonable to expect great permanent improvement from a manure so active, and which yielded go large a profit on the first crop. Yet I have seen some striking evidences of its permanency in heavy crops of clover, succeeding wheat, and in the increase of the crop of wheat on a second application. As an instance, I may mention that two years ago I sowed upon a single detached acre of "forest land," one bushel of wheat and dressed it with a barrel of African guano, costing $4, and the yield was seventeen [Pg 9] bushels. Last fall the same land, after remaining one year in clover, was again sowed with one bushel of wheat and dressed with 140 lbs. of Peruvian guano, costing $3, and the product was 22 bushels. Yet I would advise no one to rely upon guano exclusively. Its analysis shows that it contains salts of ammonia, alkaline phosphates and the other mineral elements necessary to produce the grain of wheat, but is deficient in most of the elements of the straw and roots of the plants. Hence, (says Liebig) 'a rational agriculturist, in using guano, cannot dispense with stable dung.' We should, therefore, be careful not to exhaust the soil of organic manures, but by retaining the straw of the wheat, and occasionally a crop of clover, which plant contains a large percentage of the alkaline carbonates, which are entirely wanting in Guano, furnish all the elements necessary to the entire wheat plant. In this view of the subject, and for many other reasons that I cannot stop to enumerate, there cannot be, when guano is extensively used, a more judicious rotation than the Pamunky five field system, in which clover occupies a prominent place. I have now enumerated some of the most prominent means by which you may "keep your land rich." I would not discourage the use of others. Science is daily making discoveries in the art of enriching the earth, and we should discard nothing, without a trial, which promises to be useful; always bearing in mind that the wisest economy is entirely consistent with the most liberal expenditure, in the purchase of manures, provided we take care, by judicious experiments and observation, to ascertain their efficacy, and that we get back our capital, with an actual *net* profit *in cash*, on all our investments. This latter caution is indispensable, in our country, where new lands are so abundant and cheap, that highly improved farms can never be rated in the market at their true value."

"The various manures compounded by chemists and manufacturers, should also engage your careful attention. They should not be recklessly thrown aside as humbugs, without trial or investigation, nor adopted and extensively used with blind confidence in their efficacy. I have used many of these manures by way of experiment, and the profit realized upon them has not justified me in enlarging my operations. Poudrette, manufactured in Baltimore; Bommers manure, Chappel's fertilizer and Kentish & Co.'s prepared guano, (used, it is true, upon a small scale,) have not realized the promises made in their behalf. Yet I would by no means discourage the praiseworthy efforts of the manufacturers, and hope they will persevere until, by lessening the bulk and increasing the power of their compounds, they may be able to prepare an article that for cheapness, convenience of application and efficacy, shall equal or surpass the best Peruvian guano." [Pg 10]

That desideratum, Professor Mapes believes he has already attained by the addition of superphosphate of lime to the Guano, making a compound of two-thirds of the latter to one of the former, more valuable by weight than the pure article. That being the case will greatly increase the consumption of Guano, and greatly improve the condition of all that class of farmers who desire to make their poor lands rich.

Of the use of lime, Mr. Newton has the following testimony, which we embody here for its great practical value.

"Calcareous matter is the great want of most of our lands, and in some form is essential to permanent improvement. It should be regarded as the basis of all our operations, and never to be dispensed with for any substitute. From long experience in the use of lime, I am satisfied that the French plan, of light and frequent dressings, is not only much more economical, but much safer, in our climate, than the heavy dressings common in Great Britain. Fifty bushels of slaked lime to the acre, I have found amply sufficient for any of our lands, and a greater quantity often attended with injury to the soil and crops, whilst twenty-five bushels will answer every purpose on thin lands, deficient in vegetable matter. Ashes, bone dust, and the various marine manures that abound on the shores of the Chesapeake and its tributaries, will be found important auxiliaries in the work of 'keeping your lands rich,' whilst the necessity of clover and the proper grasses, to any system of permanent improvement, is too obvious to require comment."

Although caustic lime should never be used in connection, or so as to come in contact with the Guano, there is no doubt of its being a valuable auxiliary. Upon land limed this year, Guano may be used next, and if mixed with charcoal or plaster, or plowed in and thoroughly incorporated with the soil, especially if it contains a considerable portion of clay, no loss of ammonia will occur, in consequence of the action of the lime. On the contrary, the effect will be to make the action of the Guano more active, and the immediate benefit greater; though, of course the succeeding crops would not receive as great a share. But, as Mr. Newton says, ought we to ask for great advantages to succeeding crops, from a manure which gives us such great profits from the present one.

From our notes taken upon the spot, we give a few items more in detail of Mr. Newton's operations, than he has done in the preceding quotations. The tract of land he speaks of is gently undulating; of a sandy loam, with a greater amount of clay in the subsoil; had been literally *worn out* in former years by the shallow plowing, skinning system of farming, until it would produce no more, when it was abandoned and suffered to grow up again in forest timber, principally pine of the "old field" species. No land could offer less inducements to the cultivator or give smaller hope of renovation, than these old fields of Virginia. Such [Pg 11] was the conviction of impossibility to raise a crop upon this kind of land, that Mr. Newton's first essay was looked upon by his neighbors with a conviction that the fool and his money would soon part company. One sensible old servant told us he thought his master "for sartain was done gone crazy, cause he nebber seed no nothing grow on dat land, no how could fix him." The negroes, wherever guano has been introduced, have been violently opposed to using it; not alone from its disagreeable odor and effect upon the throat and nostrils while handling it in a dry state; but because they could not be persuaded that such a small measure of stuff—200 lbs. measures about three bushels—could possibly produce any effect upon the crop. Their astonishment and consequent extravagant laudation of the effect produced, has often afforded us hours of amusement while listening to their recital of "massa's big crop," of perhaps ten bushels to the acre, which was at least double that of any one ever seen upon the same field, "fore he put dem little pinch of snuff on him."

The increase of wheat from guano may be safely calculated upon at five bushels for each hundred weight of guano used, one year with another, and up to what may be considered a fair judicious amount to be applied, which may be set down at an average of 200 lbs to the acre, upon all light soils, similar to those of that part of the country we are writing about.

GUANO vs. MANURE—EFFECTS UPON HEAVY LAND.

Mr. Newton related to us an anecdote of some value upon this point. On one of his Potomac farms, a portion of the land is exceedingly heavy—pewtery land, as it is termed from its tendency when wet to run together, presenting a glistening appearance somewhat resembling that metal. His overseer was about as unbelieving as the negroes, and declared he could beat the guano by expending the same value in manure upon a given quantity of surface. To test this and also to try its effect upon the stiff land, he applied a little short of one ton of Peruvian, which cost $50 upon ten acres, and promised a premium to the overseer if he could make a greater crop by the use of all the manure, men and teams he saw fit to apply to another ten acres lying right along side, and of the same quality of soil. Of course he spared no labor, using both lime and manure freely, but in the spring finding the appearance of his crop unequal to that guanoed, he gave it a top dressing of fine manure and a good working with the harrow. At harvest the guanoed portion was ready for the sickle several days earlier than the other, and yielded 135 bushels of a quality so very superior, it was all reserved for seed for himself and neighbors. [Pg 12]

The product of the other was 55 bushels; difference in favor of the guano, 80 bushels—8 bushels to the acre—while the value of extra manuring, probably exceeded the cost of guano, without any material advantage in the effect upon succeeding crops. In fact, it is probable, that the additional growth of straw and clover would be worth more to the next crop on the guanoed portion, than the undecomposed manure and lime would be in the other. It is needless to say both overseer and servants, were fully convinced of the virtue of guano after this experiment.

According to our notes, Mr. Newton first used guano in 1846—one ton of Ichaboe at $30, on 8 acres, with 8 bushels of seed, upon land so deadly poor, that an old negro we conversed with said; "him so done gone massa, wouldn't grow poverty grass nuff to make hen's nest for dis nigger." No attempt had been made for years to grow any crop, not even oats or rye, the last effort of expiring nature to yield sustenance to man upon one of those old worn out Virginia farms. Think of the astonishment of the poor negro, who thought his master crazy to sow wheat there *without manure*, to see 88 bushels harvested from the 8 acres.

In 1847, he used $100 worth of Patagonian upon same kind of land and reaped 330 bushels. In 1848, $200 worth of Patagonian and Chilian at $40 and $30 a ton, gave 540 bushels, which sold at $1 25, mostly for seed, on account of its superior quality. In each case the advantage to the land of equal value as to the crop. In 1849, he applied 10 tons Peruvian at $47, and 11 tons Patagonian at $30, upon 260 acres, from 75 to 250 lbs. to the acre. When we saw this crop the next spring, the appearance in favor of the Peruvian, was fully 50 per cent. upon the same cost of each kind per acre.

In 1850 he applied 30 tons, of course, all Peruvian, with equal success to former years.

Mr. Newton says, the second application of guano to the same land produces the best result—that notwithstanding the profit of the first application in the increased crop, the profit to the land is always greater.

Before leaving Mr. Newton, we will place on record one expression highly creditable to him, and convincing in its palpable truth of the value put upon this fertilizer, by a gentlemen of sound judgment and candor of speech, equal to any other within the circle of our acquaintance.

"I look upon the introduction of guano and the success attending its application to our barren lands, in the light of a special interposition of Divine Providence, to save the northern neck of Virginia from reverting entirely into its former state of wilderness and utter desolation. Until the discovery of guano—more valuable to us than the mines of California—I looked upon the possibility of renovating our soil, of ever bringing it up to a point capable of producing remunerating crops as utterly [Pg 13] hopeless. Our up-lands were all worn out, and our bottom lands fast failing, and if it had not been for guano, to revive our last hope, a few years more and the whole country must have been deserted by all who desired to increase their own wealth, or advance the cause of civilization by a profitable cultivation of the earth."

We are satisfied that the above opinion will be considered of more value—more conclusive in favor of guano, by all who are acquainted with the character of Willoughby Newton, than all else contained in the pages of this pamphlet.

OTHER WITNESSES IN VIRGINIA IN FAVOR OF GUANO.

As our principal object is to convince the skeptical, or induce unbelievers in its efficacy and value, to try experiments themselves by which they will be convinced and enriched, we offer the names of a few more gentlemen of high standing, who have been very fortunate in the use of this essential element of successful cultivation in Virginia, as witnesses, whose testimony ought to be, and will be, where they are known entirely conclusive.

Col. Robert W. Carter, of Sabine Hall, on the Rappahanock, whose land is principally of that kind of clayey loam common upon that river, once rich but badly worn by cultivation, is so well satisfied that it is profitable to make rich lands still more rich, he buys annually 30 or 40 tons of the best in market. He says he cannot afford to sow wheat without guano—it is foolish and unprofitable. He sows it broad cast, 200 lbs. to the acre, with no other preparation than breaking the lumps; plows it in; sows wheat and harrows that; in some cases has sown clover, and in others, followed wheat after wheat with increasing productiveness every year; clearly proving the effect of one application, to be beneficial to the succeeding crop. Without guano, or very high manuring, wheat will deteriorate year after year, if sown upon the same soil, until the product would not pay for the labor of sowing and harvesting.

Upon one upland field, which without manure would not pay for cultivation, he sowed one bushel of wheat and 200 lbs. Peruvian guano and made fifteen bushels. Plowed down the stubble with same application, and when we saw the crop, should have been willing to insure it at twenty-five bushels. Col. C. has nearly 2,000 acres in cultivation, which within his recollection was cultivated entirely with hoes—his grandfather would not use a plow—was as much set against that great land improver as some modern, but no more wise farmers, are against guano. Col. C. uses the best of plows; sows 200 lbs. guano to the acre and plows it in six inches deep, and sows one bushel of wheat and harrows thoroughly, but not deep enough to disturb the guano. His gain has been eight bushels ave [Pg 14] rage upon 210 lbs. guano. Thinks Peruvian at

$50 a ton preferable to any other at current prices. His land is mostly clayey loam and was so much exhausted by a hundred years hard usage, it was barely able to support the servants, until the Colonel commenced his system of improvements by draining, deep plowing, rotation of crops, lime, plaster, clover, and guano; the latter of which he looks upon as the salvation of lower Virginia; while his large sales of eight or ten hundred acres of corn and wheat, sufficiently attest its value upon that location. His actual annual profits upon the use of guano, cannot be less than two thousand dollars.

Doctor Brockenborough, Doctor Gordon, Messrs. Dobyn, Micou, Garnett and others of Tappahannock and vicinity, have all found the application even upon the bottom lands, profitable, though not to so great an extent as upon the poor old field-pine lands of Mr. Newton; but simply from the reason that his land was utterly worthless before, but after the application of the guano, was increased in value more than its whole cost, besides the profit derived from the crop.

Wm. D. Nelson, a neighbor of Mr. Newton, bought a tract of land for a residence, at $4 an acre, which in its natural condition was not worth cultivating; but with guano will pay all expenses of that and the cultivation and the cost of the land the first crop.

Upon a portion of this land, a poor sandy loam, he applied 200 lbs. Peruvian guano and one bushel of wheat per acre, and made 12 bushels, while a strip through the field, purposely left without guano, did not produce the seed, and remained as destitute of clover as though it never had been sown, forming a very striking contrast to the luxuriant growth upon each side. In another trial he made 10 bushels from one sowed, with 200 lbs. of Patagonian guano, of a very good quality. This is about in proportion to the current price of the two kinds, though the latter cannot be so certainly depended upon for good quality as the Peruvian. Another trial was made with 1,100 lbs. Peruvian and 1,100 lbs. Patagonian, and 11 bushels of seed upon 11 acres which made 160 bushels of wheat of very fine quality, and large growth of straw. Upon 36 acres, same kind of soil, well manured in the previous crop of corn, sowed 36 bushels and made 162. The first had not been manured. The evidence in favor of guano in this case, needs no comment. By an outlay of $40, a much more valuable crop was made from the 11 acres than from the 36; the permanent improvement to the land from guano was much greater than from the manure. In this case the guano was plowed in about four inches deep.

Mr. Nelson thinks the yield of wheat will average in that neighborhood, an increase of 16 bushels for 200 lbs. of Peruvian guano. [Pg 15]

H. Chandler, Westmoreland Court House, bought a farm at a price for the whole below the cost of the mansion house alone, because the land was so utterly and hopelessly worn out, as to be past the ability of supporting those engaged in its tillage. When we saw it, we should have been willing to insure the growing crop of wheat at 20 bushels, the result of 210 lbs. of Peruvian guano to the acre; while the clover upon the stubble of the previous year could not be excelled in point of luxuriousness upon the richest field in the State of New York, where the land was valued at $100 an acre.

Mr. Chandler first commenced with 250 lbs. African guano, measuring 3½ bushels, to the acre, upon which he sowed one bushel of wheat. The result 17 bushels to the acre upon land which only gave 5½ bushels in any previous crop. Cost of guano $5; profit, $6 50. The next year he gained an increase of 12 bushels to the acre over previous years, by the use of 250 lbs of Patagonian guano; while the clover, Mr. Chandler thinks, worth more than the whole cost of the application. A still better result was produced last year from 210 lbs. of Peruvian. The soil is a yellow clayey loam, which in its unimproved condition looks about as unpromising for a crop, as the middle of a hard beaten road.

Mr. C. tried guano upon river bottom land, but the improvement was not so remarkable.

We were assured by Mr. C., that many persons who had long been accustomed to look upon the hopeless barrenness of this land, were wont to stop as they rode past this field of clover, and look at it with utter astonishment. Some could not be satisfied with looking, but would drive to the house to inquire what magical power had been used to produce such a strange metamorphosis in the appearance of the place. When assured it was all effected by guano, they went away — not satisfied — but unbelieving.

What tends much to increase the effect of this improvement, is the fact, that directly opposite lies another tract, still in its barren condition, lately purchased by Dr. Spence, a very enterprising gentleman, imbued with the spirit of improvement, which will soon be brought into the same condition, notwithstanding its unforbidding appearance.

Mr. S. B. Atwell who owns an adjoining farm, has been equally successful in the use of guano. Before using it, his wheat upon 20 acres was hardly sufficient to pay for harvesting. The first crop after using it, 400 bushels. He has also increased the crop of corn from 20 to 260 barrels by lime, guano and clover. In the meantime, the land has increased in value in about the same ratio.

In Lancaster County, we saw a field of wheat on the farm of Dr. Leland, sown upon corn ground, one part with 200 lbs. of Peruvian guano [Pg 16] to the acre, the other with a full dressing of hog-pen manure, by the side of which the ground was seen in its natural barrenness, scarcely making a show of greenness; while the rank growth of the guanoed portion made as great a contrast with that manured upon the opposite side.

Guanoed wheat upon the farm of Col. Downing in the same county showed as great a contrast with land both limed and manured; while directly alongside of this luxuriant growth, the land was as destitute of vegetation as a brick pavement.

The effect of guano upon strawberries, Col. D. found to excel anything else ever tried.

A neighbor of Col. Downing had a fine show for a wheat crop on exceedingly poor land from the application of only 90 lbs. Peruvian Guano to the acre.

Capt Wm. Harding, Northumberland, C. H., assured us he made 27 bushels per acre upon only tolerably fair land, by the use of 200 lbs. Peruvian guano, plowed in and followed by clover, worth more than the guano cost.

Col. Richard A. Claybrook, in the same neighborhood, made 15 bushels—the land along side almost as bare as the surface of the guano islands.

We might mention a dozen others in the same place, in fact in most of the places mentioned, whose testimony would be as strong as those we have named.

Col. Edward Tayloe of King George Co., having been very successful in the use of guano, induced his neighbor, Wm. Roy Mason, Esq. to test its powers by the most severe experiment we have ever known it subjected to. He selected a point of a hill, from which every particle of soil had been washed away, until nothing in the world would grow there. It would not produce, said he, a peck of wheat to the acre, but with a dressing of 300 lbs. African guano, it gave me thirteen bushels, and now while that is covered with clover, other, so called, rich parts of the field are almost bare. A field which had never produced for years, over four bushels of wheat to the acre, was dressed with 250 lbs. of guano and one bushel of plaster at a cost of $7 to the acre, which gave thirteen bushels of a quality greatly improved, and a very large growth of straw, which he esteems highly as a top dressing for the clover, which far exceeded upon the guanoed land that which was highly manured. The success of Mr. Mason was so flattering, he immediately purchased six tons for the next experiment.

If all the faithless would pursue the course indicated in the following *experiment with guano*, by Mr. Richard Rouzee of Essex Co. Va., they would probably be as well convinced as he, that the greatest "humbugging" about guano, is in neglecting to profit by its use. He says:—"I [Pg 17] must confess that I have been skeptical in relation to the various accounts of the fertilizing properties of guano, especially in these times of humbuggery, and therefore determined to subject it to the most rigid test." In view of this, on the 3d of October last, I selected two acres of land by actual measurement, proverbially poor, never having yielded in a course of ten years cultivation more than three bushels per acre, and in consequence, was called by way of derision, "Old Kentuck." To the two acres 560 lbs. of guano were applied in the most injudicious manner by strewing it on the top of the corn bed—the consequence was, when the wheat was ploughed in, and came up, a small girth was only seen on the top and a space between each row at least one third of its width; in this condition it remained until about the middle of November, when it had so sensibly disappeared, that it attracted the attention of one of my neighbors, who remarked to me, that at least one half of it had been destroyed, in which opinion I concurred; in examining that which remained, we were of opinion that three-fourths of it had from three to ten flies in the maggot state on each stalk; in this state of things I surrendered all hope of any tolerable return, more especially as the rust made its appearance in it a short time before it ripened. — Now for the result —

The 2 acres of land yielded me 32 1/4 bushels of wheat at $1 per bushel,	$32 25
Deduct for average yield of the above, 2 acres, 6 bushels at $1 per bushel,	$6 00
Deduct for Cost of 560 lbs. Guano,	$12 70
	– – $18 70
	$13 55
Add for additional straw,	50
Clear profit,	$14.05

Here is a clear profit of $14 upon $12.70 invested, and acknowledged to be applied in the most injudicious manner. It is easy to judge what would have been the profit under different circumstances. In the vicinity of this city where straw sells for $5 per hundred little bundles, instead of a credit of 50 cents it would have been at least half the cost of the guano.

GUANO IN NORTH CAROLINA.

Henry K. Burgwyn's first trial with guano. Its effect on grass sown with wheat. — The name and farm of this gentleman is so widely known as a successful renovator of miserably poor worn out fields, that we are delighted to have it in our power to have his testimony to our impregnable [Pg 18] array of witnesses in favor of the most valuable substance for the improvement of such land, ever given by an overruling power for the benefit of those who ought to be exceedingly thankful for so good a gift. But hear what this writer has to say upon this interesting subject.

"Having about 150 acres of my wheat, this year sown upon last year's corn ground, and the land being rather light and not too rich, I feared lest I should fail with my grass sown on *this* wheat, because of the two successive cereal crops; I therefore bought guano, mixed it with its bulk of plaster, then added fine charcoal, the same, and to this mixture double the whole bulk of deposit of the Roanoke river, a rich alluvial earth, and sowed the whole broadcast in February and March, and harrowed it in, on the top of the wheat I sowed at the rate of 200 lbs. of guano to the acre; the value of which, no doubt, was doubled by the mixture with the absorbents of the ammonia, which is so exceedingly volatile even when left for a few hours, is easily dissipated by the March winds. On this land, I had sown in October previous, clover, timothy, Kentucky blue grass, and Italian ray grass. My harvest has now been over, three weeks, and I have never had a finer stand of all these, even on our rich bottoms. The ray grass matured its seed, rather sooner than the wheat was two-thirds as tall, and where *very thickly sown*, materially injured the product of the wheat, *I have reaped an increased product from my wheat, amply sufficient to repay my outlay for the guano, plaster, &c., and have my grass as my profit on the investment*; this in turn will shade and improve my land, fatten my stock, increase my crops, and cheer my eye with 'grassy slopes,' in place of 'galled hill sides;' this is profit sufficient for the most greedy if turned to a proper account; — be it remembered, too, this was a light and rather poor soil, but based on a good clay subsoil."

To this we beg leave to add from our own knowledge of this land, which is situated on the Roanoke river 6 or 7 miles below Halifax, that it was before being improved by Mr. Burgwyn, about as unpromising a tract as can be found upon all the "cottoned to death," poor old fields of that sadly abused State. In the condition it was when we first saw it, while undergoing the operation of putting a four horse plow through the broom straw and old field pines, notwithstanding our strong faith in the ability of such men as the Messrs. Burgwyns to redeem such land from its condition of utter and apparently hopeless barrenness, we must own, that if Mr. B. had made the assertion while we were riding over this very tract, that within two years he would reap a remunerating crop of wheat from the barren waste, and coat the ground with a carpet of luxuriant grass, we should have told him the day of miracles had passed away. But we had not then seen as much as we have since of the miraculous power of Peruvian guano. [Pg 19]

We might continue to cite hundreds of similar cases but propose to pass over into Maryland, and after showing its application there has produced equally beneficial results, travel northward, calling here and there a witness as we proceed. Among others, we may call to the stand in Maryland, will be the editor of the American Farmer, whose testimony we consider almost invaluable, having devoted much attention to the subject, and to whom, and his able correspondents, we desire to award full credit, in this general manner, to save repetition, for much of the information we shall give the readers of several of the succeeding pages. The testimony of witnesses of such high standing, cannot be too highly estimated by those who are anxious to learn how to renovate their worn out farms, or make the rich ones richer.

EXPERIMENTS IN MARYLAND.

Effects of guano upon the crop to which it is applied. — Edward Stabler, in the American Farmer, thus speaks of an experiment he made in 1845, soon after the introduction of guano to any extent into this country.

"In a field of some 10 acres, one acre was selected near the middle, and extending through the field, so as to embrace any difference of soil, should there be any. On this acre 200 lbs. of Peruvian guano, at a cost of about $5 was sown with the wheat. Adjoining the guano on one side, was manure from the barn yard, at the rate of 25 cart loads to the acre; and on the opposite side (separated by an open drain the whole distance;) ground bones were applied on the balance of the field, at a cost of $6 to the acre; the field equally limed two years preceding. There was no material difference in the time or manner of seeding; except that the manure was lightly cross-ploughed in, and the guano and bones harrowed in with the wheat.

"The yield on the guanoed acre was 35 bushels; the adjoining acre with bone, as near as could be estimated by dozens, and compared with the guano, was about 27 bushels; and the manured, about 24 bushels. The season was unusually dry; and the manured portion suffered more from this cause than either of the others; the land being considerably more elevated, and a south exposure."

In our opinion Mr. S. is in error in regard to the manured land suffering most from drouth. In our experience we have always found the best effects from Guano, in wet seasons, or upon irrigated land. He says also, "This is one of the most active of all manures; and although he thinks the effect evanescent, it might aid materially in renovating worn out lands." Since that time a great many other Maryland farmers have, undoubtedly come to the same conclusion, for notwithstanding the price, which he thinks too high to justify its extensive use, has not been ma [Pg 20] terially reduced, there is more guano sold in Baltimore than any, or perhaps all the ports in the United States; and the benefits derived from its use upon the worn out lands of Maryland, have been of the most satisfactory character.

In speaking of the after crop of grass upon the land above mentioned, he says:

"The field has since been mowed three times; the first crop of grass was evidently in favor of the boned part; the second, and third, were fully two to one over the guano, and also yielding much heavier crops of clover seed. On a part of one land, 18 bushels to the acre of the finest of the bone were used; on this, the wheat was as heavy as on the guanoed, and the grass generally lodges before harvest, as it also does on much of the adjoining land with 12 bushels of bone."

This is all right; it should never be mixed with lime, and it should be plowed in. In his experiments, the lime in the soil had the effect to disengage the ammonia, and not being sufficiently buried or mixed up with the earth to prevent its escape during a very dry season, much of its value went afloat in the atmosphere. If he had given a bushel of plaster as a top dressing, there is no doubt the effect upon the grass crop would have been entirely different. The action of guano is very variable upon different soils, as well as upon the same kinds of soil in different seasons, or from the different manner of applying it; but there is one thing in its favor, it seldom fails to pay for itself, as Mr. Newton remarks, in the first crop; and if properly applied, that is, plowed in with wheat, upon poor, sandy, "worn out land," and followed by clover, and that dressed with plaster, it will pay far better in the succeeding years than the first. This has been fully proved in a hundred cases, since Mr. Stabler tried his experiments; for two years after, in writing upon the same subject, he says "Harrowing in the guano with the wheat will generally produce a better crop; but its fertilizing properties are more evanescent. I prefer plowing it in for all field crops; and when attainable, would always use it in conjunction with ground bones, for the benefit of succeeding grass crops. This is pre-supposing that you determine to improve more land than the resources of the farm will accomplish, and are willing to do it by the aid of foreign manures; and being 'far removed from lime.' If the object is to realize the most in a single crop, and to obtain the quickest return for the outlay, use the guano alone, and harrow it in with the wheat; but the land, according to my experience, will derive but little benefit from the application, unless the amount is large. By plowing it in, particularly if mixed with one third its bulk of plaster, the effect is decidedly more durable; nor is it then necessary that the seeding should so immediately follow its application. If, however, the object is to improve the land at the same time; [Pg 21] and surely it should be a primary object with every tiller of the soil—and lime, from your location, or the price, is unattainable, I would advise about half the amount determined on, to be expended for ground bones. This may be harrowed in with the wheat."

It is surprising what an effect a few bushels of ground bones to the acre will produce; reference is made to a single experiment, and not an isolated one either. Some six years since, we applied ten to twelve bushels of coarsely ground bones to the acre, on about half of a twelve acre field; on two lands adjoining, was guano, at the rate of 200 pounds to the acre, (the cost of each about the same,) and extending nearly through the field; both were applied in the spring, on the oat crop—and which was decidedly better, by the eye, on the two lands with guano. In the fall, the field was sown with wheat, manuring heavily from the barn yard, adjoining the guano, but not spread on the two lands, or on the boned portion of the field.

There was but little difference perceived in the wheat, except from the manure, which was the best—the field having been limed for the preceding corn crop, 80 bushels to the acre. The experiment was made to test the comparative durability of the three kinds of manure; the guano, ground bones, and manure from the barn yard; and the ultimate profit to be derived from each, in a full rotation. After the first crop of grass, and perhaps the second, which was in favor of the manured portion, the succeeding crops of hay and clover seed, have been decidedly better on the boned part of the field. At the present time, and also the past season, this being the fourth year in grass, the guanoed lands present about the same appearance, that does a small adjoining space, purposely left without manure of any kind, lime excepted. The manured part affords good pasture, but is quite inferior to the boned, which would give a fair crop of hay, and probably three times as much grass as the two lands with guano. It is believed that the increased crop of clover seed on the boned, over the guanoed portion, paid for the former; and that the two crops of clover since taken from the field, have paid, or nearly so, for the lime or other manures applied.

This evidence corresponds with the opinion of Professor Mapes; that is, that the value of an application of guano is greatly enhanced by the addition of phosphate of lime, in some shape; the guano acting immediately and producing a direct profit, while the slow action, for which some farmers cannot wait, keeps up the fertility for years, or until the owner may find time to profit by another application of guano.

We quote again a few more of the very sensible remarks of friend Stabler. "I am an advocate for the liberal use of all kinds of manure, guano included, if the price will justify it. A farmer had better buy manure than to buy grain, if compelled to do either; for we cannot ex [Pg 22] pect much from nothing, or reasonably calculate upon improving very poor land without manure of some description, unless plaster will act with effect; nor is this generally the case without the land possesses naturally, some particular source of fertility, not wholly exhausted by bad or improvident tillage.

"It is probable those will be disappointed who expect to do everything with guano—make fine crops and improve the land, while they take everything off, and dispense almost, if not entirely, with the more permanent manures, all equally within their reach. True, we may exist for a time, only half fed and half clothed; but it is just as reasonable to expect to improve under such a regimen, as to calculate upon continued, not to say increased fertility of the soil, without an ample supply, of the right kind of manure.

"With all its acknowledged advantages, it may be questioned whether there is not one drawback to the introduction of guano. It is used with less profit in direct connexion with lime, than with most kinds of manure; and its facility of application, and quick return, has induced many to give up the lime entirely, if not also to some extent, to neglect the resources of the farm. Others again, in improving poor land, advise the guano first, and the lime afterwards. This may do very well; but is often better in theory than in practice, for the lime is omitted altogether, and perhaps at some risk of loss, in both time and money, as regards permanent improvement. To use a figure of speech—the prudent architect will first secure a solid foundation to build upon, and with materials of known durability; this accomplished, he need have no fears of the stability of the structure, and may, at pleasure add thereto, either for ornament or utility."

"That thin lands may be brought to a very productive state, by the liberal and repeated applications of guano, there is no doubt; but at what cost and how durable the improvements might be, I am not prepared to say. In two instances, from 700 to 800 lbs. were applied at one time to an acre; but in neither did the results correspond with the expense, or induce a repetition of the experiment. My own experience so far, is in favor of more limited applications, say 100 to 200 lbs. to the acre, (taking in consideration the price of both grain and guano,) and also used in connection with other manures, which is found to be the most profitable, and probably more durable in its effect; in two experiments, with from 50 to 150 lbs. of guano to the acre applied three years since with barnyard manure, for wheat, the effect on the grass crop at this time, is quite marked; applied in this way, it hastens maturity — thus, in a degree, guarding against rust — renders the grain more perfect, and is believed to be one of the most profitable modes of using guano." [Pg 23]

Nothing could be more sensible than the advice of this gentleman, not to rely upon guano alone. To waste or neglect stable and home made manures, or throw away bones or other valuable fertilizers, because we could buy guano, would be as insensible as it would for a man to throw away a handful of bank bills, because he happened to have just then a pocket full of gold and silver coin.

We never have, nor shall we recommend guano to the exclusion of everything else; but we do recommend every farmer in America, to whom an additional quantity of manure would be an object, to buy guano; because he will be almost sure to derive a certain and immediate profit from the investment. It will make poor lands rich, and rich lands richer.

EVIDENCE OF THE DURABLE EFFECTS OF GUANO.

Upon this point, we have the following testimony of Thomas P. Stabler, of Montgomery County, Md., a gentleman of the highest degree of intelligence and integrity; one of the society of Friends, who are rather noted for not being extravagant in their expressions or encomiums of an article, without good grounds therefor. We make these remarks, because, as every good lawyer will tell you, the character and standing of your witnesses is of more importance than their language, to make a strong impression in your favor.

In speaking of the means within reach of farmers, by which they can renovate their worn out lands, of which Maryland has an ample share, friend Stabler says, "In some districts the distance from lime is so great, that the man with small means can scarcely be expected to use it upon a large scale—but in regions of country where bone, guano and poudrette act favorably, none need be without important aid from their use. Under a judicious system of cultivation and correct management, either of these will make bountiful returns the first year, and the strongest and most conclusive evidence exists of their durability as manures. Proofs of this abound in my neighborhood. Reference to the 'facts' in a single case in point may suffice for an example. In the summer of 1845, I prepared seventeen acres and a few perches of land for wheat About five sixths of this was extremely poor—upon a portion of the field, was put 112 ox-cart loads of manure from the barn yard and stable, on what I considered about an average quality of the land. On the 12th of the 9th month, (September,) I sowed seven bushels of wheat on this part of the ground and plowed the manure and wheat in together with the double shovel plow—very soon after the balance was sowed with 270 pounds of good African guano per acre, for which I paid $40 per ton, and plowed this in with the wheat, immediately after sowing, [Pg 24] in the same manner as the other. During the succeeding winter and spring, the appearance of my wheat field became the subject of much notice and remark on the part of my neighbors, as well as others from several adjoining counties who saw it, many of whom supposed that this application of guano could not possibly produce such a crop as its then present appearance indicated—in this, however, they were disappointed—there were two small pieces left without manure of any kind. One of these upon the best part of the field, and the other upon a part of medium quality.

"It may be recollected that the crop of wheat that season was generally most inferior, both in quality and quantity. Upon the parts left without manure, it was scarcely worth cutting, and men of integrity and good judgment, were of the opinion that without the aid of the guano, I could not have saved more than 60 or 70 bushels of wheat from the field. The product was 320 bushels, that weighed 64 lbs. to the bushel. The guanoed portion continued at harvest to be decidedly better than that manured from the barn yard and stable. This field was sown with clover in the spring of 1846, and to this time its appearance affords as strong evidence of great improvement in the land, as it did during the growth of wheat. It has now been pastured freely during two summers, and been exposed to the action of the frosts of two winters, and upon the guanoed portion I have not yet seen a single clover root thrown out of the ground, while from the part manured from the barn yard, it has almost entirely disappeared. Good farmers have frequently remarked during the present summer that the appearance of this field warrants the conclusion that it is now capable of producing largely of any crop common to our country.

"Thus 'worn out land' is renovated, and ample means produced for increasing its fertility. Similar instances of improvement exist in very many examples that can be seen in this portion of our country, resulting from the application of lime, bone and poudrette, as well as from guano."

Guano prevents clover from being thrown out by frost.—We wish to call back the attention of the reader to this reliable statement of Mr. Stabler, not only for its importance to farmers, but because the same thing has been remarked by other gentlemen who have used guano. It can only be accounted for from the fact, that guano seems to be peculiarly adapted, more than any other manure, to give the young clover a vigorous start, so that in its early stages it acquires a growth too strong to be affected by the usual course of freezing and thawing, by which less vigorous plants are thrown out. For this reason alone, if guano had no other value, farmers in some sections of the country where the soil is [Pg 25] peculiarly affected by this difficulty, would find their account in the use of an article which would enable them to grow clover, for clover is manure, and it should be a sine qua non with every farmer to avail himself of all the means within his reach to increase the supply of manure from the products of his farm. Let him not depend alone upon the purchase of guano, but rather upon the means which that brings within his reach of increasing his home supply by the growth of clover, and largely increased production of straw. Those who are interested pecuniarily, which the writer is not, in the increased sale of guano in the United States, have no fears that our recommendations to make manure at home—to use lime, plaster, bones, clover, and every other source of fertility within their reach, will decrease the sale of guano. On the contrary, those who are most disposed to use all these sources of fertility, are the very men most disposed to use a substance which all experience has proved superior to all others. Besides, there is, and probably always will be, enough "worn out lands" which can be profitably renovated, to use up all the guano which will ever find its way into this country. So our earnest recommendation is, where lime is available, let no man claiming the honorable title of farmer, fail to make the application. Let him also gather up all the fragments—let nothing be lost—make all the manure at home he possibly can, and then he will not only have the means, but a disposition also to buy that which a beneficent Providence sends him from the coast of Peru; of the good effect of which we will prove by further testimony—that of the Hon. James A. Pearce, Senator from Maryland, and a farmer of no small note in that State. He says—"In April 1845, I applied 350 lbs., probably of African or Patagonian guano to an acre of growing wheat, the land being entirely unimproved and very poor. It was applied as a top dressing, of course, but mixed with plaster." (In what proportion he does not say, but we will by and bye; but he does say)—"*The wheat was doubled in quantity at least*—fine clover succeeded it—and in two crops, one of corn and one of small grain, three and four years afterwards, the effects are still apparent." Now this effect was produced by the use of the guano as a top dressing; a method universally acknowledged to be the most unfavorable to the development of the full value of the application.

The editor of the Farmer in answer to an inquiry whether a combination of charcoal, plaster, and guano would make a profitable *top dressing* in spring for wheat, says, "yes"—but thinks if it had been plowed in with the seed in the fall, the result would have been much better. However, says he, "we entertain not the slightest doubt, that, if his wheat field be top dressed with the mixture next spring, it will greatly increase the yield of his wheat crop, unless the season should prove a very dry one, as the charcoal, and plaster, will each tend to prevent the escape [Pg 26] of the ammoniacal gases of the guano, and as it were, offer them up as food to the wheat plants.

"In April 1845, I applied 350 lbs. of guano to an acre of growing wheat, the land being entirely unimproved and very poor. Of course it was applied as a top-dressing, *mixed, however, with plaster*. The wheat was doubled in quantity at least; fine clover succeeded it; and in two crops, one of corn, and the other of small grain, last year and the present, the effects are still apparent."

If our correspondent would *mix*, in the proportion of 200 lbs. of *guano*, one bushel of *charcoal*, and half a bushel of plaster per acre, and sow the mixture on his wheat field next spring, after the frost is entirely out of the ground, then seed each acre with clover seed, and roll his land, we have no doubt that his wheat crop would be increased five or six bushels to the acre, perhaps more, and that he would have a good stand of clover plants, and a luxuriant crop of the latter next year.

"Our opinion is, that *guanoed* land should always be sowed to clover, or clover and orchard grass."

In this, particularly the opinion of the last paragraph, we fully concur—to obtain the full value of guano it must either be mixed with plaster or charcoal, or what is better, plowed in and thoroughly incorporated with the soil, and the land always sown with clover, peas or some other plant of equal value for green manure. It is true Col. Carter has been successful with wheat after wheat; while many continue successful, by carefully retaining all the straw; the guano being sufficient to keep up the everlasting ability of the soil to produce an annual crop of grain.

THE FIVE FIELD SYSTEM AND GUANO.

We look upon this as the most preferable of all other systems of farming ever adopted in the South—it is the system of Edmund Ruffin, to whom Virginia owes a debt of gratitude beyond her power to pay. It will be seen from the following extract from a letter of Mr. Newton that that eminent agriculturist is of opinion that improvement of poor land is unlimited, if guano in connection with this system is perseveringly applied. He says—"The "five field System," which is now rapidly extending over all the poor and worn lands that are now under improvement by marl, lime, or guano, originated, or at least was first extensively introduced in lower Virginia, on the Pamunkey, and has there wrought wonders, aided by marl and judicious farming. The rotation is corn,—wheat,—clover—wheat, or clover fallow,—and pasture, and after pasture one year, commencing the round again with corn. This system, if guano be applied to both crops of wheat, on corn land and fallow, or alter [Pg 27] nately with lime or marl, when calcareous manures are required, will readily increase the crops and permanent improvement of the land. In the commencement of the rotation, lime had better be applied with the putrescent manures to the corn crop, to be followed by guano on wheat. If this system be perseveringly, pursued, I can scarcely see any reasonable limits to the improvement of poor lands and the increase of the profits of agriculture."

Disappointment will result from the application of lime, marl, salt potash, guano, or any special and highly concentrated substance as a fertilizer, to the neglect of organic manures. We lay down this fact as incontrovertible, that no soil, however fertile it may be made for the time being by any of these special manures, can remain permanently so, unless care is used to maintain a healthful supply of organic matter,—rich mould—good soil upon the land cultivated. If this is done, we never shall hear of guano failing to bring increased crops or of the "land running out," where it has been applied. Special manures of any kind may fail to produce crops, where this essential requisite to good farming is neglected. Guano, in our opinion, should always be followed by crops of clover, grass, peas, or some crop that will shade the earth, and can be turned under with the plow, to keep up the necessary supply of nitrogenous food for cereal crops.

The effect of Lime and Salt upon land is to *dissolve* the inert portions of organic matters in the soil, so that plants can suck up their substance into their own composition. Both are highly beneficial, but insufficient to add permanent fertility.

The effect of guano, is greater than any other highly concentrated manure ever discovered and applied to any soil. Its benefits are immediate continuous, and unlike lime, without exhausting the soil of its organic matter. Yet its benefits will be increased by the addition of organic manures derived from green crops, straw, or the stable, and the value of these will be greatly increased by the addition of lime, salt and plaster, while any deficiency of phosphates must be supplied by powdered bones or another application of guano.

The effect of plaster with guano is to arrest the excursive disposition of the volatile parts of the guano, and imprison them in the earth until called forth by the growing plants to do the State some service. The following question to the Editor of the American Farmer, and his reply, are to the point in this matter:—

A correspondent says—"As to the question of mixing plaster with guano, there is one question I should like to propose to the editor, viz.—'what will be the effect of sowing guano upon land by itself, and then, the seed being in the ground, giving it a heavy top-dressing of plaster, so as to arrest the 'excursion,' of which so much is said?" [Pg 28]

Reply by the editor.—"The effect of such application of guano and plaster would be, to prevent the waste of the ammonia of the former, as every rain would decompose more or less of the plaster, separate the sulphuric acid from the lime, and the sulphuric acid when liberated, would unite with the ammonia, form a sulphate of ammonia, and hold the latter in reserve to be taken up by the roots of the plants. The presence of plaster with all *organic* manures, either directly mixed with them, or broadcasted after they may be applied, tends to prevent the escape of their volatile parts. We prefer them together for two reasons,—*first*, because, by bringing the two into *immediate contact*, the action of the plaster is more direct;

and *secondly*, because the time and expense of one sowing is thereby saved. We go for saving every way, as time and labor costs money, and we look upon economy as a virtue, which should be practised by all, and especially by husbandmen."

If the plaster and guano is mixed together, 25 lbs. of the former to 100 lbs. of the latter, will be found a proper proportion, and sufficient to prevent the ammonia from making an "excursion." Unless the soil be very poor, 200 lbs. of good Peruvian guano is as much as we should recommend for wheat. In this we have the concurrence of the editor of the Farmer, and perhaps a hundred gentlemen whom we have conversed with upon this subject. All agree in the opinion, whether mixed with plaster or not, that a judicious application of guano will more certainly restore productiveness to worn out land, or add fertility to that already productive, than any other substance ever applied.

Want of Faith in the efficacy of guano.—Whatever doubts may have existed in the minds of careful men, there is no room for doubts now, that Peruvian guano possesses regenerating properties beyond belief, without evidence, and capacity to increase the productiveness of lands in sound condition, in such an eminent degree, that any farmer who has the power to obtain it, evinces great folly and perverse obstinacy, if he continue to cultivate his land without applying it; either for want of faith, or pretended disbelief in its efficacy; or because he thinks the price fixed upon it by the Peruvian Government, "unjustifiably high;" or because although he has no doubt it will answer in the moist climate of England, is sure it will never answer in this dry climate; or because he is afraid the luxuriant crops produced by the application of guano will exhaust his land; or because his neighbor Jones killed all his seed corn by putting only a handful in the hill; while Mrs. Jones killed all her flowers and fifty kinds of roses with the "pisen stuff;" and therefore he don't want any more to do with it; or because it has failed to give remuneration under the most injudicious application, made contrary to all instructions or experience of those who have used it; or for any and all the other thousand and one objections raised by those who have ne [Pg 29] ver used it, and seem determined they never will; probably because when the almost miraculous accounts of its operations were first published, they had cried out "humbug" so loudly they are determined no after evidence shall convince them the only humbug in the case was in their own disbelief. It is for the benefit of these unbelievers we are now writing. Our object is to present such an array of facts guaranteed by such respectable names, they shall have no hook to hang a doubt upon—no reason—no justifiable excuse for any sane man longer to neglect to apply an article of such positive, certain benefit to his hungry soil.

ED. REYNOLDS ESQ., OF BALTIMORE, ON THE VALUE OF GUANO.

Writing on the subject of "bought manures," as everything is termed not produced upon the farm, and how dubiously they are looked upon by some persons calling themselves good farmers, for fear of being humbugged, Mr. Reynolds says, in a letter dated July, 1850, "Since 1843, I have been trying to find out which is the best of all these 'new things,' and have now, after having been very considerably humbugged, settled down upon bones and guano—although, even the last named in a very dry year, has also 'cheated me'; but this is by no means its character, as I am constrained to admit, that after having tried it on all sorts of soil, and perhaps as long if not longer than any other person in the State, it is my opinion that when properly applied, with an average fair season, it is a very powerful fertilizer. My mode of using it is, when applied to tobacco, to mix one and a half bushels of the Peruvian, (which is ordinarily 100 lbs.) with one bushel rich earth, and one bushel of plaster, which admits about the fifth part of a gill of the mixture to each hill for every 5,000 hills—and putting it in the center of the check before being scraped—so that when the hill is made, it lies beneath the plant. On wheat, I apply three bushels of Peruvian guano equal to 200 lbs. mixed with one bushel of plaster, one bushel rich earth to the acre, sowing on the surface and plowing it in as soon and as deep as possible, after it is sowed. The past spring I have put 300 lbs. to the acre, on 30 acres of corn, being half of a field, on a farm in Calvert, mixing with it the same quantity of rich earth and plaster, and sowing on the surface, plowing in at once very deep, using the cultivator only in working it afterwards. I do not intend to use it at all with corn, hereafter, but not because I do not think it also a good fertilizer with this crop, (as my corn on my Calvert farm, upon which it has been used, now shows very fair,) but only because it has never failed to pay me three fold better on wheat, than on anything else. In order to test its virtue, it is essentially necessary to plow it in deeply, and stir it as little as possible afterwards. [Pg 30] "

Bones.—Of these I have used both ground and crushed, and always to advantage at ten to twelve bushels per acre; bought from manufacturers here, and agents of houses in New York; but I am using the crushed dissolved by oil of vitriol, as prepared by myself on my farm in Calvert in the following way: The bones, (which we buy in the neighborhood at 50 cents per 112 lbs.) after breaking them with a small sledge hammer on an old anvil, we put at the rate of three bushels in half a hogshead, and apply to that quantity 75 lbs. oil of vitriol, filling up the half hogshead to within eight inches of the top with water, letting them remain, (but stir the contents occasionally with a stick,) say two to five weeks, according to the quality and strength of the vitriol; then start the contents of the half hogshead into a large iron kettle, apply a slight fire and the whole contents will in less than an hour be reduced to a perfect jelly. We use two half hogsheads at once, to prepare it expeditiously. We then mix the contents of each kettle, with a horse cart load of rich earth, or ashes, throwing in a half barrel of plaster, mix or compost it handsomely, and use at pleasure, on an acre of land with any crop you choose, and you will have permanently improved two acres at the following cost, viz: Bones, $1.50, vitriol, $3.75, plaster, $1.12, making $6.37, or $3.18 per acre, and this may be repeated so as with proper attention, as much lasting improvement may be made each year as many farmers derive from their barn yards. Bones in any form never fails to show their striking effects on clover and other grasses—but either bones or guano will scarcely ever fail to produce a better crop of clover, which, with the increased quantity of straw, (particularly when guano is used,) will enable and encourage the saving of larger quantities of barn yard manure, and which must inevitably cause a lasting improvement.

This coincides with our views exactly, as we have in all these pages endeavored to impress upon our readers, that the increased growth of straw from the use of guano, will increase the manure pile, and "inevitably cause a lasting improvement."

Poudrette.—"I have used also, to good advantage, particularly on clayey lands, at the rate of six to eight barrels per acre. It is a first rate top dressing on young clover in spring, at two to three barrels per acre; this article has been prepared so badly heretofore, that a great quantity of it was really worthless."

We also concede to poudrette as much credit as Mr. Reynolds but as will be seen, it will cost more to improve land with it than with guano.

Prepared Guano — Agricultural Salts — Generators and Regenerators.— Of these, the testimony of Mr. Reynolds is exactly to the point, concise and strong, and exactly in accordance with all the facts we have been able to collect upon the same subject. He says, "I have tried them on corn, wheat, oats, clover and tobacco; but have yet to discover that they [Pg 31] ever generated anything for me, though I have heard them sometimes well spoken of."

Want of room in this pamphlet alone prevents us from inserting the names and operations of many other gentlemen in this rapidly improving State—a State now undergoing the process of renovation by the use of guano, to a greater extent, perhaps, than any other in the Union.

GUANO IN DELAWARE.

Hon. John M. Clayton's Farm. — No one who looks upon this highly improved farm now, with its most luxuriant crops, can be made to believe it was a barren waste seven years ago — hardly worth fencing or cultivating. This great change, so far beyond the power of human belief, has been effected by lime, plaster and guano. The railroad from Frenchtown to New Castle, passes through this farm, four miles from the latter place. It is well worthy a visit from any one anxious to make personal observations of the effects of "bought manures," upon a soil too poor to support a goose per acre.

Effect of Guano on Oats. — During a visit to Mr. Clayton, in 1851, we saw the most luxuriant growth of oats upon one of the fields of this farm, which we have ever witnessed, and it has been our fortune to see some tall specimens of this crop on the bottom lands of Ohio, Indiana, and Illinois. The seed he had obtained from England, and the means of making it grow, from Peru. The guano was plowed in with the oats, at the rate of 350 lbs. to the acre. The soil is a yellow clayey loam. The effect upon other crops had been equally beneficial. The growth of clover was so great he had purchased thirty bullocks to fatten, for the purpose of trying to consume some of his surplus feed. The effect upon wheat, corn, potatoes, turnips, garden vegetables and fruit trees, was almost as astonishing as upon the oats and grass.

C. P. Holcomb, Esq., one of the most improving farmers of one of the most improving counties in the U.S., has met with great success in the use of lime, plaster, and guano. His beautiful highly improved home farm is near Newcastle; but that upon which he has met with great success in the use of guano, lies about four miles from Dover. Before he purchased it had become celebrated for its miserable poverty. It is now equally celebrated for its productiveness. The use of guano in that part of the State has now reached a point far beyond what the most sanguine would have dared to predict four years ago; and the benefits are of the most flattering kind. Lands have been increased in value to a far greater extent than all the money paid for guano; while the increased profit from the annual crops, has produced corresponding improvements in the condition and happiness of the people.

No greater blessing, said an intelligent gentleman to me, ever was bestowed upon the people of Delaware. [Pg 32]

Extensive use of Guano by a Delaware farmer. Maj. Jones, whose name is extensively known as a very enterprising farmer, purchased in the summer of 1851, of Messrs. A.B. Allen & Co. New York, sixty tons of Peruvian guano, for his own use. With this he dressed 300 acres of wheat, upon the farm at his residence on the Bohemia manor; plowing in part of it and putting in part of it by a drilling machine at the rate of 200 lbs. to the acre, sowing the wheat all in drills. Part of the ground was clover, part corn, and perhaps one half wheat and oat stubble. The earth at the time of sowing was so dry, doubts were entertained whether it would ever vegetate; and that and other causes extended the work so late, upon a portion of the ground, there was scarcely any appearance of greenness when it froze up. With all these disadvantages, the crop was estimated at harvest at twenty bushels to the acre. Without guano no one acquainted with the farm would have estimated the crop at an average of ten bushels. This gives an undoubted increase of five bushels for each hundred weight of guano; and as the soil contains a good deal of clay with which the guano was well mixed, it will retain much of the value of the application, for the next crop. Maj. Jones has heretofore derived very great benefits from the use of guano, as might safely be adjudged from the fact of his risking $3,000 in one purchase of the same article.

Lasting effects of Guano. — Maj. Jones is well satisfied upon this point. In 1847, he used 16 tons, half Peruvian and half Patagonian, sowed with a lime-spreading machine and plowed in deep, say eight inches on clayey loam — planted corn and made 60 bushels per acre on 100 acres; which was an increase of 12 bushels per acre over any former year. Next spring the weeds grew as high as his head on horseback. Rolled them down and plowed under and sowed wheat, five pecks to the acre, and made a heavier crop than ever before made on same land, which he attributes entirely to the guano. Thinks the third crop of wheat is benefitted from guano plowed in three years previous.

The extent to which guano is used in the State of Delaware may be inferred from the fact that it is not at all unusual for merchants in small country villages to purchase from 50 to 200 tons at a time for their retail trade.

Among other successful users of guano in that State, we may mention Governor Ross, who, if as good a ruler as he is farmer, ought to be continued in office to the end of life.

The soil to which guano has been mostly applied in this State is a sandy loam, and the process of applying it, by sowing broadcast from 200 to 350 lbs. per acre, and plowing in from four to six inches deep, previous to sowing wheat, which is always followed by clover, by every one who understands his own true interest; for wherever that course has [Pg 33] been pursued, there has been a certain profit derived from the application, even when the wheat has failed.

The improvements in farming in Delaware within the last ten years, will probably exceed in proportion to acres and people, any other State in the Union. Nearly all the northern part of the State has been whitened with lime, and the southern part is rapidly following the same path; while the sale of guano in all parts will exceed any other section of the country, if not in quantity, certainly in numbers of persons making use of this sure means of restoring the lands of an almost ruined State, to their pristine fertility.

GUANO IN PENSYLVANIA.

There has probably been less guano used in this great State, than in her little sister, of which we have just been speaking. This may be owing to the fact that great improvements have been made by the use of lime, and that Pensylvania farmers generally are not much inclined to leave the path their fathers trod before them; or that they are skeptical as to what they hear of the miraculous powers of guano; hence, its use has been in a great measure confined to market gardeners, or experiments in a small way; the sales at Philadelphia, for home consumption, so far as we have noticed, are mostly in small lots of one to ten bags. Among all with whom we have conversed, however, who have used Peruvian guano in that State, we have never heard a doubt expressed of its value, though the idea, strangely enough seems to prevail, that it will only be profitable for gardners and small farmers, and that it is of no benefit to succeeding crops. No doubt the progress of improvement by the use of guano in that vicinity has been greatly retarded, in consequence of the sale of considerable quantities of "cheap guano," which however low in the scale of prices, is still lower in the scale of values. In fact, there is but one thing connected with the spurious stuff, lower in any scale, and that is the honesty of those who manufacture or knowingly sell such a villainous compound to farmers, who are utterly ignorant upon the subject, under solemn assurances, that it "is equal to any guano in market, and only a little more than half price."

Mr. Landreth, the celebrated seedsman of Philadelphia, applied $500 worth of Peruvian guano last spring, principally on the bean crop—he thinks guano admirably adapted to all the Brassica tribe, including turnips, cabbages, rutubaga, radishes and all cruciform plants. Upon a lawn which appeared to be running out, he applied guano, and the grass is now green and vigorous. The character of his soil may be judged from its location; it is on the Delaware river above Bristol, and had [Pg 34] been awfully skinned before he came in possession. Now, with a liberal expenditure for manures, he gets two crops a year.

Guano for grass lands.—The Germantown Telegraph says: "The application of guano broadcast to grass lands has been found to produce a decided difference in the crop. In several instances this season, where Peruvian guano has been applied at the rate of 200 lbs. per acre, about the middle of April, the yield of hay has been double in quantity, over the intermediate lands not so treated; and in every instance noticed, it is believed that the difference in quantity produced will amply repay the cost of the guano."

GUANO IN NEW JERSEY.

Guano has not been extensively used in New Jersey, owing to the abundance of green sand marl, which is a very valuable fertilizer, abounding in that part of the State most in need of artificial manures. Guano has, wherever used, produced the most astonishing results. One of these we witnessed upon the farm of Mr. Edward Harris, a gentleman well known for his enterprising spirit of improvement and intelligence in agriculture, who resides at Moorestown, which lies in the sandy region east of Philadelphia. He sowed 400 lbs. to the acre, plowed in with double plow, sowed oats and seeded with timothy, which upon similar soil often "burns out" for want of shade, after the oats are harvested. Not so in this case. The shattered oats from a remarkably fine crop, vegetated and grew with such a dark green luxuriance, there was more danger of the young grass being smothered out; so he had to put the mowers at work, who cut heavy swaths of this second crop of oats, for hay. If it had been situated so it could have been fed off, the amount of pasture would have been almost incalculable. It is needless to say the effect of guano upon this land, was not evanescent. Other trials made by Mr. Harris, have convinced him of its value to Jersey farmers, and that good as "Squankum marl" undoubtedly is, farmers would do better to expend part, at least, of their money in guano.

The name of James Buckalew is known, perhaps, more extensively than any other in New Jersey, as one of her most enterprising, rapidly improving, money making farmers, whose testimony in favor of guano may be easily obtained by any one who will take the trouble to go and see what beautiful farms he has made out of the barren sands near the Jamestown station, on the Camden & Amboy railroad, by the use of lime, plaster, marl, manure and guano. It is a pity that every one who doubts the feasibility of profitably improving the worst land in that State, by the power of such an agent as Peruvian guano, could not see what has been done by Mr. Buckalew. Let them also look at what were [Pg 35] once bare sand hills around the residence of Commodore Stevens, at South Amboy, a gentleman who ought to be more renowned for his improvements on land than water, notwithstanding his world wide reputation, in connection with the yacht America. Go ask how it is that these drifted sand hills have been covered with rank grass, clover, corn, turnips and other luxuriant crops; the very echo of the question will be, guano.

Look at the astonishing crops of Professor Mapes, at Newark. Peruvian guano, in combination with his improved superphosphate of lime, hath wrought the miracle, aided as it has been, by the deepest plowing ever done in that State.

Mr. Samuel Allen, at Morristown, has now growing upon a poor barren, gravelly knoll, a crop of corn which might put to blush the owner of a rich and well manured field, and which ought to put to blush some of the unbelievers in the power of guano to produce such a growth upon such a soil; rather where there was no soil, hardly enough to grow a respectable crop of mullen stalks. Mr. Allen has tried guano for several years upon every kind of garden vegetable, with the most wonderful success. A crop of Lima beans now growing exhibit its wonderful power in the strongest manner. The application has been made by a small dose at planting and two sprinklings hoed in during their growth.

A great many other persons in this State have produced most wonderful effects upon land almost utterly worthless, while in the immediate benefits, those who have applied it to lands in good condition, have profited more than with double the cost of manure.

Guano for Peach Trees. — A New Jersey nurseryman assured us of his firm conviction in the power of guano to cure the yellows in peach trees — that no grub or worm can be found alive in the roots of a tree where guano is applied — that young trees can be brought into bearing by the use of guano, a year earlier than by any other forcing process with which he is acquainted.

GUANO ON LONG ISLAND. [1]

One gentleman assures us he tried an experiment very carefully, and found an application of guano at two and a half cents a pound, 300 lbs. to the acre, more economical than hauling his own manure one mile. The fair value of team work and cost of labor hired, was more to the acre than the guano, and the first crop quite inferior, the second no difference, and the third slightly in favor of the manure. He thinks buying city manure, particularly street sweepings, about the poorest use to which he could put his money, as he certainly could make 50 per ct. more upon the same amount expended in Peruvian guano. Professor Mapes [Pg 36] entertains the same opinion, about hauling manure, where guano, or rather with him, guano improved by the addition of his "improved superphosphate of lime," can be procured.

Dr. Peck, a gentleman well known for his philanthropic motives in settling and improving the "Long Island barrens," has proved that every acre of that long neglected, and until quite recently considered worthless portion of the Island, can be rendered fertile, so as to be cultivated with great profit, either in farms or market gardens, by the aid of this greatest blessing ever bestowed by Providence upon an unfertile land.

Several of the Messrs. Smith, of Smithtown, could show any Long Island farmer who still has doubts upon the subject, that guano is the greatest worker of miracles in this age — that it is just as capable of producing great crops on the barren sands of the Island, as it is on the tide water shores of Virginia, upon soil of the same character.

A great deal has been said in deprecation of the waste of fertilizing matters in the city of New York, in which the writer of this pamphlet has conscientiously joined; because, he thought it wicked to commit such waste, while we were surrounded by lands lying idle, for the want of these very substances. Precious, however, as they would be to the farmer, he cannot afford to use them. That is, it would be poor economy for a Long Island farmer, no matter how near the city, to expend money in the hire of men, vessels and teams, to save, carry, haul and apply to his farm, the immense amount of fertilizing substances now wasted; because the same capital expended in purchasing and applying guano, will produce a much greater profit. The difference in cartage is enough to astonish one who has never thought upon the subject. One man with a pair of horses can easily carry guano enough in one day, thirty miles into the country, to manure ten acres of ground. To carry an equivalent of city manure, in the same time, would require 300 pair of horses and 350 men. Who can wonder that barren lands have remained barren? Who will not wonder if they still continue so, with such fertilizers as their owners might possess to render them otherwise? But few of the residents in the interior of Long Island, if the manure was given to them, can afford the time and team work to haul 300 loads for ten acres, while all can afford the time for one load; and they may be morally certain the capital invested in that load will be returned in the first crop. The great advantage of guano over all other manures is, the concentration of immense fertilizing power in such small bulk.

Guano in New York and Connecticut, generally, has been less used than any sound reason will justify. A comparatively small portion of the market gardeners—a few gentlemen in the improvement of rural homes, and here and there a nurseryman, have derived immense benefits; but the bulk of the farmers are still either faithless, or ignorant; in most [Pg 37] cases the latter, of the benefits they might derive from a liberal expenditure in the means, and the only means within their reach, of rendering their lands productive.

Effect of Guano on Garden Seeds.—From the society of Shakers, at Lebanon, so justly celebrated for growing garden seeds, we receive the most positive assurance that no manure ever applied by them, has had such an effect as guano. The production of seeds of all descriptions, is not only increased, but the quality is improved to an astonishing degree. The same effect has been noted upon wheat, particularly in our account of Mr. Newton's operations. So also has it in England. This view of the case should give an additional value to guano to the farmer, as not only an improver of the quantity of his products, but by the gradual improvement in the quality of the seed, calculated to be of vast benefit to him in that respect. Garden seeds raised by guano, as soon as their superiority becomes known, will be in such demand that no other can be sold. Another advantage will arise from the fact that such seeds will be found entirely free from weeds, as none grow after a few years upon land manured only with guano.

The beautiful residence of Mr. Edwin Bartlett, near Tarrytown, exhibits strong evidence of the fertilizing power of guano upon the poor, unproductive hill sides of Westchester Co. That place, now so luxuriant, was noted a few years ago, as too poor to support grasshoppers. It was the poverty stricken joke of the neighborhood.

[1] For interesting letters from Long Island, see appendix.

GUANO IN MASSACHUSETTS.

We have heard a good many assertions that guano, however valuable it might be upon the warm sandy soils of the south, would not answer in the cold land and climate of the New England States. To refute this fallacy, we have some strong testimony. Seven years ago, while the very name of guano, and much more its virtues were unknown to half the farmers of America, Mr. S. S. Teschemacher, of Boston, a gentleman of science and practical skill in gardening, became so fully convinced of its value to the cultivators of American soil, he published a pamphlet for the purpose of inducing others to profit by its use. From that pamphlet we make a few extracts. He says—"One of the numerous objections to this manure is, that, although it may answer well in the humid atmosphere of England, it cannot produce equal benefit in the hot, sandy soils of this country. In reply to this, it may be observed, that the sandy soils of South America are more hot than they are here; and, on the coast of Peru, where it is most used, it scarcely ever rains at all. The truth is, that it certainly requires moisture to decompose it, and enable it to enter into the juices of the plant; by no means, however, so much [Pg 38] as is usually supposed; but, once absorbed by the roots and plants, it imparts that strength and solidity which enable them to resist both drought and cold.

"It is beyond dispute that guano contains the chief ingredients required for the growth of plants. The instances hereafter adduced will show that the combination and form of these ingredients are such as to promote not only its immediate action, but clearly to accelerate considerably the progress of vegetation."

The chief ingredients, then, of guano, are,

Ammonia, in various forms and combinations;
Phosphate and oxalate of lime and magnesia;
Salts of potash and soda;
Animal organic matter;
Sand and moisture.

Besides the evidence we have given of the value of an application of such a compound, it contains evidence within itself to every mind embued with any knowledge of agricultural chemistry, that it will not only promote immediate growth of vegetation, but produce a lasting benefit to the soil. It contains all the materials necessary for the growth of cereal or esculent vegetation in the exact form required—that is an impalpable powder—to promote rapid, certain, large growth, and abundant fruitfulness, and consequent profit.

EXPERIMENTS BY MR. TESCHEMACHER.

To Indian corn, applied one teaspoonful to the hill, well mixed with earth, at time of planting. When twelve or fifteen inches high, hoed in three tea spoons full around the corn, and covered two inches deep and watered. Soil—a poor, sandy, sterile one. Product—one seed produced three main stalks with eight perfect ears and five suckers, weighing 8¼ lbs. The best plant without guano, weighed 1¼ lbs. and only had one ear.—"I find the best mode of applying guano is to hollow out the hill, put in one teaspoonful and a half of guano, and mix it well with the soil. Spread even, then put on this about one or one and a half inch depth of light soil, on which sow the seed and cover up. When the corn is about twelve inches high, or the time of first hoeing, begin with the hoe about four inches from the stems, and make a trench the width of the hoe about two or three inches deep. Spread in this trench about three or four teaspoonfuls guano, stir it in, and cover the trench as quickly as possible. If this last operation can be performed just before or during rain, the action will be quicker and more effectual."

Four or five teaspoonfuls of dry powder producing such an effect, is what staggers the belief of those who see with their own eyes. [Pg 39]

So great is the luxuriance of growth from such an insignificant application, it is necessary to increase the space nearly double between the hills. In a country where fodder is so valuable as it is in Massachusetts, the great increase of stalks is of equal importance with the increase of grain. Indian corn requires both phosphate of lime and magnesia which it finds in guano, in combination with ammonia, in a state just ready to be absorbed by the growing plant, wherever brought in contact, with its roots.

Mr. T. found the guanoed corn planted May 22d, ripened sooner than that planted May 1st. with manure. This alone on account of the difficulty from frost, is sufficient to give it great claim upon northern farmers.

Effect on Grass.—"The application of this manure to grass land already laid down is for many reasons often attended with uncertain results. The best mode is, to spread broadcast about 250 lbs. per acre of the Peruvian guano as soon as the snow is off the ground. It would be very advantageous if, after it was spread on, some light loam could be put over it, in the manner of a top dressing. I state the Peruvian guano is the best for this operation, as it contains what Dr. Ure calls *potential ammonia*, or ammonia in a more permanent form; whereas the ammonia from the Ichaboe guano evaporates more easily, and this valuable ingredient is therefore lost in the atmosphere when it is spread on the surface.

"Most excellent crops have been obtained, where the grass is sown and laid down in the autumn, on light, sandy soils, by sowing the guano evenly broadcast, then harrowing twice, sowing the grass seed, and rolling."

The best mode of applying it, however, is to sow broadcast and plow it in—at the south, on sandy soils, no matter how deep—at the north on soils more clayey, plow it in about four inches deep—the real object being to so mix it with the soil as to prevent the escape of ammonia, which is exceedingly volatile. Remember, *Guano* should never be used as a top dressing, except in combination with plaster, or some other substance which will prevent the escape of the most valuable portion of its composition.

In several case, where sods have been laid down for lawns or embankments round houses, the most surprising growth has been obtained by strewing the surface with guano previous to laying on the sod.

E. Baylies, of Taunton, sowed 460 lbs. African guano per acre, with grass seed, which yielded, this year, one ton per acre more than that without; and the appearance of the guanoed grass is now much more thick, luxuriant, and promising, for next year than the other.

"Another friend of mine sowed grass in sandy soil with a full quan [Pg 40] tity of manure, and an adjoining acre, with 400 lbs. Ichaboe guano. The guanoed acre grew stronger, and retained its full verdure the whole winter; the manured piece, on the contrary, became, as usual, brown by the action of the frost."

Mr. T. as well as nearly all the English writers upon the subject, has noticed the improvement in quality as well as quantity of grain and garden vegetables. It is a well authenticated fact, that birds wont touch the manured wheat, while they can obtain that which is much more plump and rich where guano has been applied.

Effects on Trees and Grape Vines.—"The experiments with guano on trees which have come under my observation, including exotics number about one hundred and fifty. The action has invariably been to produce large foliage, of a deep healthy green."

The best mode of applying guano to fruit-trees, or flowering shrubs, is to dig it into the earth at such distance from the trunk as will be likely to meet the largest number of fibrous roots.

"For instance, round an apple-tree of ten years' standing, dig a trench one or one and a half foot deep, at about the same distance from the stem that the branches extend; let this trench be about one foot wide; then put at the bottom one and a half inch depth of guano, dig it well in, and incorporate it with the soil; then cover up carefully and press the earth down. The effect of this application will unquestionably be felt for several years."

On grape vines, the action of guano has been proved exceedingly beneficial; increasing the growth of vines and fruit, improving the flavor and hastening the ripening, so as to escape early frosts.

In planting young trees, put about a pint in the bottom of the hole covering with soil so the roots will not touch it. No insects or grubs will disturb the roots of such a tree.

"Several friends, who have tried guano this year on their pear-trees, have reported to me the result to be greater crops, and of a much larger size, than they ever had previously."

Guano on Peas—Method of Applying.—The kinds on which I experimented were Prince Albert, Shilling's early grotto, (a dwarf pea,) blue imperial, and marrowfat. Draw a deep trench with a hoe, strew guano in the trench, mix it up with the soil, over this put about one inch and a half of earth, then sow the seed, and cover up. The quantity used should about equal the quantity of seed. The produce of the three first kinds of peas, was five full pecks to the quart of seed, besides a full quart of seed gathered for next year. From the marrowfats I obtained only four pecks and a half, and no seed. The growth of all was extremely luxuriant. The marrowfats were six and a half feet high, the stems from one to one and a quarter inch in circumference. Guano [Pg 41] should be placed at such a depth that the natural moisture of the earth will decompose it and render it fit for the plant. In the lightest soils—plow and bury guano a little deeper than in others more heavy; the guano itself retains moisture, and absorbs it naturally.

Guano on Beans, doubled the yield of a paralel row, while the improved flavor was perceptible to those who had no idea of the cause which produced it. In drouth, the power given plants by guano, to resist the scorching rays of the sun, is remarkable.

On Melons, the effect was equally favorable, giving a large increase of highly flavored fruit.

On Potatoes.—We give out of many equally favorable, only one experiment, just to show the ability of farmers to grow this crop in the most unsuitable soil, by a small expenditure for guano, twenty per cent. better than with manure. Here it is. "Soil, very sandy and light; quantity, 800 lbs. African (per ship Samos) to the acre; cost, $20. Same soil, with twenty-two loads fine compost manure, cost $22. Yield, as eleven to nine, or twenty-two per cent. in favor of guano, the potatoes with which were larger than the others."

On Turnips, no manure is equal to guano. The crop has been doubled in numerous instances. Mr. T. says of one experiment he made, "The plants on this portion are now twice as large as those which have not had any. It is perfectly beautiful to see the luxuriance of all these guanoed vegetables compared with the others."

On Strawberries, nothing has ever been applied equal to guano, provided the plants are plentifully watered. The best mode of application is in solution. One pound is enough for ten gallons of water.

On Cauliflowers.—Two experiments, one with guano, the other with a solution. The first are fine strong plants, particularly one to which I gave a larger share than the other; it is heading finely. But those with the solution are much larger and finer. I have been accustomed to observe the cultivation of this vegetable, and never saw such a luxuriant growth. They are now, (Sept. 15th) beginning to show flower; and, if the season is favorable, I expect the heads will be very fine. The plants are at least four times larger than those on the same piece without guano, or any manure at all, planted on the same day, from the same seed bed.

On Rhubarb or Pie Plant, guano has the most decided beneficial effect, increasing the size, flavor and tenderness of the stalk; besides the very great advantage of bringing it forward some two or three weeks earlier in the spring. Fork it in all over the bed, just as early as the frost will permit, at the rate of 600 lbs. to the acre.

On Asparagus, the same treatment will more than double the quantity of this excellent, healthy vegetable. In the fall, give a dressing of salt [Pg 42] equal to 15 or 20 bushels to the acre. With the guano, nothing else need be applied, if it is thoroughly mixed with the soil.

For Vegetables, Plants, Trees, and Shrubbery generally, where fruit is an object, apply the guano as above, in powder. Where flowers of rare size and beauty are desired, apply it in solution, or by frequently stirring in small dressings just before a shower. Another important observation on this subject is, that guano, or its solution, should never be applied except at that period of the season when the growth of wood is proper and natural.

In forcing houses, nothing can be equal to guano. One thing, it produces no weeds, or insects; this is enough to insure its favor wherever it may be tried.

On roses, the beneficial effect is already well known. If tea roses are cut down when the bloom is over, repotted in fresh earth, and well watered twice or thrice a week, with guano water, they will immediately throw out luxuriant shoots, and be covered with their fragrant blossoms. The cactus tribe will bear a larger quantity and stronger solution of guano, without injury, than most other plants.

"During the progress of my experiments," says Mr. T., "I have been delighted with the unfailing and extraordinary luxuriance of growth and produce on a miserable spot of land, induced by the use of this manure, and struck with the numerous instances which have come to my knowledge of erroneous applications of it. On a stiff clay, guano would be of little value, except on the surface, or an inch or two deep, unless it were considerably lightened by the addition of sand, or well broken up by exposure, in ridges, to frost, as every clay soil should be. A light, porous, sandy soil would require 300 lbs. Peruvian, or 400 lbs. best Ichaboe; and for this soil I think the Peruvian best adapted, as it retains the ammonia longer, and, being less soluble in water than the Ichaboe, its qualities are not so soon washed out."

In a soil already much enriched with manure, and at the same time abounding in phosphate of lime, I have found the guano to produce less visible effects than on a poor, sandy soil.

Most excellent effects have been produced by steeping seeds in guano water of moderate strength for eight to twelve hours, dependent on the kind of seeds, and then planting with one to three inches soil between the seed and the guano. The steep encourages the growth of the young plant, whose roots, in a more advanced stage, find the guano, which continues the stimulus.

Quantity for a Steep.—Put one, one and a half, or two teaspoonfuls of guano, according to quality, in a quart bottle, shake up, and when settled, use; then refill and use two or three times, previous to putting in fresh guano. Or, in the large way, from fifteen to twenty gallons of [Pg 43] water to one pound; mix in a barrel, stir up and leave it to settle, taking care, however, to put a cover on, to prevent the escape of ammonia.

DIRECTIONS AS TO QUANTITY AND MANNER OF APPLYING GUANO TO VARIOUS CROPS AND SOILS.

The best action of guano is undoubtedly upon naturally poor or worn out light sandy soils. Next sandy loam—then loam proper—then clayey loam or exhausted gravelly soil, and lastly cold stiff clay, or land naturally wet. Upon the first particularly at the south, it should always be plowed in from four to six inches deep; and will always afford the greatest profit when applied to wheat land and that sown with clover.

Preparation of guano for use.—Until some ingenious Yankee invents a cheap mill by which he will make a fortune and the lumps be easily ground, the following method may be pursued. Take the bags on the barn floor or in some close room with tight floor and sift the guano over a box, through a 3/8 mesh sieve, putting the fine back in the bags and lumps on the floor. These may be mashed with a stout hoe or shovel, or with a block like a pavier's rammer. Sift and break again until all is fine. Lay the dust with a very slight sprinkle from the nose of a watering pot; of a solution of copperas, at the rate of 10 lbs. to the cwt. of guano, or with plaster or loamy earth—woods mould or dry fine clay. Many persons prefer to mix plaster with the guano in the first instance at the rate of a peck of plaster to a bushel of guano—others use an equal weight of each. Where plaster is not to be had, from five to ten bushels of pulverized charcoal or dust from the coal pit, or pulverized peat, to each hundred weight of guano may be used to fix the ammonia and prevent loss. Sulphuric acid 1 lb. to 10 of water, with which to sprinkle the mass may be used as a fixer. But if it is kept in the bags, in a dry room, until ready for use, and then prepared, sown and plowed in at once with as little exposure to the air as possible, very little of the ammonia will escape. The true axiom to be observed in the use of guano, is to plow it in as soon as possible after it is sown and before it is moistened with dew or rain; and to plow it in deep, or in some way thoroughly incorporate it with the soil, so that rains will not wash it away, or hot sunshine cause it to evaporate. We hold all top-dressings with guano, to be wasteful, on account of its volatile character, and because it needs the moisture in the earth to fit the substance of which it is composed so its fertilizing properties can be taken up by the roots of the plants. If spread upon the surface, it must wait for a dissolving shower to carry it down to the roots; in the meantime, it is moistened by dews and evaporated by the sun, and carried off to enrich your neighbor's crops half as much as your own. [Pg 44]

Preparing Land and Sowing. — When ready to plow the land for wheat, measure an acre and lay it off in lands 18 feet wide; put the guano in a pail and walk up one side and down the other with a moderate step throwing handfulls across at each step, and you will find you do not vary much from two hundred pounds to the acre. Never sow in a windy day if it can be avoided, nor faster than it can be plowed in the same day.

To prevent guano from getting into the mouth and nostrils. — Take a thin piece of sponge and wet it and tie over the mouth and nose. Whenever the dust accumulates, wash it out. If you must sow while the wind is blowing, mix earth enough with guano to prevent blowing away.

Depth it should be plowed in. — On light sandy land, there is no danger of its ever being plowed in too deep. On sandy loam, it ought to be plowed under at least six inches — eight inches would be better. On true loam, a less depth will answer, though we are strong advocates of deep plowing. On clayey loam, four inches will answer, and on clay, particularly in the Northern States, if well harrowed or put in with the cultivator, there will be no great loss of ammonia, as the clay is a great absorber of that volatile substance. This rule may in general be observed; upon the light lands of the south, it cannot be too deeply buried; in the clay lands, or in the more heavy, cold, or moist lands of the north, it may be covered too deep to benefit the first crop; but, if the after cultivation is good, whatever is planted will be sure to be benefitted. Upon granite soils, it will be of less value than silicious or aluminous ones. Though most valuable on poor sandy or worn out old fields like those of Virginia, already described, still it must not be rejected by the owner of any land which can be improved by manure, because this is a manure of the very best and most concentrated kind; containing more of the ingredients necessary to promote vegetable growth, in the exact proportion and combination, ready prepared for use, than any other substance in the known world. It is a fertilizing substance which none will reject who once learn its value, unless very deeply prejudiced. It is idle to reject it because the Peruvian Government wont let us have it at our own price, because we can profit by it at theirs. It is nonsense to say, it will answer in the moist climate of England, but not in our dry one. Truth deduced from experience, in several States, in various climates and soils, refutes all such sayings. Besides, it has been used with continued success in the burning sun and soils of Peru, ever since the conquest by the Spaniards, and, according to tradition for ages untold previous to that time.

Guano on Wheat.—We repeat, sow broadcast and plow in upon all light lands, *deep*; at the rate of 200 to 600 lbs. to the acre, as you can afford, or as the land requires—we believe in the small quantity and re [Pg 45] peat the next sowing, to be by far the most judicious. On heavy lands you may harrow or cultivate it in, but the plow is better. It will do well on lands previously limed, but should never be mixed with lime or ashes, unless mixed with plaster or charcoal. If you must use it as a top dressing in the spring, mix a bushel of plaster with every hundred of guano, sow and harrow in—don't be afraid of injuring the wheat Always sow clover or grass on guanoed grain.

On Indian Corn.—Follow the same directions as for wheat, or if the land is already rich, and you wish to give the corn an early start, scatter at the rate of 100 to 200 lbs. guano in the furrow, and cover it two inches deep with another furrow and then drill the corn. Be sure and never let the seed come in contact with the guano, or you will kill it most certainly. Guanoed corn should be sowed in wheat, particularly whenever it has been dressed with a large quantity.

To growing Corn, if it is desirable to apply it, turn a furrow away from the row on each side and scatter in the bottom at the rate of 300 lbs. per acre, and turn back the earth immediately.

Green Corn—roasting ears—are improved in taste by guano beyond anything ever conceived of by the lovers of this luscious food.

Quantity per acre.—Thomas S. Pleasants of Petersburg, Va., a well-known writer upon agriculture, and who has had much practical experience ever since the first introduction of guano into this country, says: — "*Corn* is a gross feeder and will take up a greater quantity of guano than perhaps any other crop. I have known as much as 600 lbs. applied to the acre and the product was in proportion. Each hundred pounds will give an average product of ten bushels as various experiments have proved From the above mentioned application of 600 lbs. a product of 73 bushels was obtained, which left 13 bushels as the product of the soil alone. For corn, guano may be spread broadcast on the land and ploughed in as deeply as it is desirable to break the soil; or it may be strewed along deep furrows to be afterwards ridged over and the cultivation to be in only one direction. The best result I ever obtained was from this latter mode, when from land not capable of producing five bushels, I harvested a crop that could not have been less than 35 bushels to the acre.

"The furrows were opened deep and wide by passing the plow both ways and the guano strewed along these at the rate 1 lb. per every ten yards. They were then covered over and the land thereby thrown into beds. But in whatever way it is used, the roots of the corn will be sure to find it all, and between these two modes, I think there is little or no choice. I would certainly advise against putting it in the hill, though I have sometimes seen good results. It is difficult, however, in such a case, to prevent the guano and seed from coming into close contact; [Pg 46] and, unless there are two or three inches of earth interposed between them the seed will be certainly destroyed."

For wheat, the guano should be spread broadcast at the time of seeding the wheat, at the rate of 200 lbs. to 250 lbs. per acre and ploughed in. If the land has been previously fallowed, it will be sufficient to plow it in with a one horse plow; if broken up for the first time, there will be no objection to using a larger plough. The best depth for getting it in, however, is, I think, from four to six inches. It always acts more powerfully on clean land; indeed if there is much crude vegetable matter in the soil, there is frequently little or no advantage derived from its application. Experience, therefore goes to show that the most economical application is to corn land; that is, to land that has just produced a crop of corn, no matter how poor it may be. If it is intended to be put on land that has been lying in grass, it would be advisable to fallow it as early in the season as practicable, and afterwards to get it in with a small plow as already suggested.

The same direction will apply to oats and also to rye. But for oats, 125 to 150 lbs of guano will be as much as can be used to advantage.

A. B. Allen of New York, one of the earliest, and most strenuous advocates of using guano, who, long before he ever thought of being engaged in its sale, used to distribute small parcels among farmers and gardeners to enable them to try experiments and learn its value, in a letter to the Southern Cultivator, says:—"Never put guano in the hill with corn, no matter if covered two or three inches deep; for the roots will be certain to find it, and so sure as they touch the guano, so caustic is it, that it will certainly kill the corn; the same with peas, beans, melon vines, in fact most vegetable crops. Wheat and other small grains have so many roots, and tiller so well, there is no danger of guano killing them, when sown directly with the seed. Still, as before remarked, it is better to plough it in before sowing the seed.

"After corn is up, you may apply a table spoonful, at the first time hoeing; dig it an inch or two deep six inches from each stalk. A table spoonful to the hill will take 250 to 350 lbs., per acre, according to the distance the hills are apart. If the soil be rather poor, a second dose at the time the corn first shows its silk, will add considerably to the yield in grain, if followed by rains, but little or nothing to the growth of stalk. Guano increases the size of grain more than stalks; hence one must be content to wait till the grain is fully matured before giving an opinion of the virtues of guano.

"Before applying the guano, it is better to mix it well with an equal quantity of plaster of Paris or charcoal dust. Either of these substances help to retain the ammonia and prevent its evaporation. [Pg 47]

"The genuine unadulterated Peruvian guano, is so much superior to any other kind, it is in reality the *cheapest*, though the price is considerable higher than that of the other qualities."

Guano on Oats.—Mr. Allen says, "I am satisfied from experience and observation in the use of guano, for the past twelve years, that the best method, decidedly, of applying it to crops in our dry climate, is to plow or spade it into the ground; and autumn is the best time for doing this, as it gives time for the pungent salts contained in the guano, to get thoroughly mixed with the soil before spring planting. Do not fear to loose the guano by plowing it in as deep as you please—it will not run away, depend upon it. At the south, it loses half its virtue if not plowed in at least three inches deep; six or twelve inches would be still better."

Because "autumn is, for many reasons, the best season" for applying guano, as a general thing, we do not recommend an application to this crop, notwithstanding our full conviction it will increase the product upon any light, poor soil, from ten to twenty bushels to the acre, for each cwt. applied. As some however, will find it more convenient and profitable to manure the oat than wheat crop, we recommend them to plow in from 200 to 300 lbs. to the acre, on ground that was clean tilled the previous year, and sow the oats in drills, three or four bushels to the acre and seed with clover, herds, or ray grass. If not to be followed with grass, we would use a much less quantity; say 125 or 150 lbs. to the acre. As may be seen in the account of Mr. Harris' crop, not one half of the 400 lbs. was taken up by the oats. With wheat, on the contrary, the guano is dissolved more slowly by winter rains, giving the crop a vigorous growth in fall, and sometimes all winter, so it sends out double the number of stalks in spring. The sun too, is so much less powerful at that season, evaporation does not take place so easily as in summer.

Great Crops from Guano.—In England, 48 bushels of wheat and 100 of oats have been made from an acre dressed with 200 lbs. of guano. A late English writer, in detailing his own experiments, and urging others to the same course, says; "The reason guano is serviceable to all plants arises from its containing every saline and organic matter required as food. It is used beneficially on all soils; for, as it contains every element necessary to plants, it is independent of the quality of the soil. So far as the experiments in England and Scotland may be adduced, one cwt. of guano is equal to about five tons of farm-yard manure, on an average; but it is much higher for turnips than for grass."

Guano on Grass.—As we are opposed to using it as a top dressing, of course we shall not recommend its application to this crop. Generally, by using it on wheat and other crops, the farmer will save manure enough to top dress his meadows. Nevertheless, in combination with [Pg 48] proper ingredients, we do say it is a good and profitable manure for grass. For each acre mix from 200 to 400 lbs. with as many bushels of plaster, or ten to one of charcoal, or twenty to one of dry swamp muck or peat, woods mould or fine clay, and sow upon the meadow or pasture early in spring. If the season is moist, the benefit will be very great; if dry, it will probably be said, as it has been before; "Oh, this guano is good for nothing—I tried it once on grass and it never done a bit of good."

On potatoes, 400 lbs. to the acre, broadcast, may be used to good advantage, if it is plowed in deep enough, on clean land. As it is a caustic manure, and requires a good deal of moisture, as well as potatoes, it is not suitable for the hill or surface dressing. A less quantity will pay a greater profit to the immediate crop, without much after benefit, if it is drilled in the bottom of a deep furrow and then covered by turning two furrows, one from each side, so as to leave a slight depression between them, and directly over the guano. Upon these beds plant the tubers in drills. After hoeing, scatter a mixture of equal parts of lime, salt, ashes and plaster, a large handful every yard, all over the rows, and we will warrant the crop free from the potato rot.

On turnips, nothing can exceed guano, unless the phosphate of lime in bones could be rendered equally pulverulent. Use 3 to 600 lbs. per acre, and plow it in at the last plowing, and top dress with five bushels of ashes and two of salt as soon as the turnips are up. Follow with wheat or rye and grass. One half the above quantity and five bushels of bone dust dissolved in sulphuric acid, will produce a wonderful crop of turnips, or ruta bagas. Guano may be used to equal advantage upon all kinds of root crops.

Benefits to the Dairy Farmer.—The beneficial use of guano in the manufacture of butter and cheese, is unquestionable. In many districts in England, and in some in this country, the continual cropping of grass and conversion of it into cheese, has so exhausted the soil of its phosphates, the milk will no longer produce the quantity of casein necessary to make cheese making profitable. When this is the case, you will find the cows seeking to supply the deficiency by eating bones. Wherever guano has been used upon pasture land, it is found that cows eat the increased luxuriant grass most greedily, and improve not only in quantity but quality of their milk. We cannot, therefore, recommend too earnestly, to all dairy farmers, to give their pasture lands an immediate dressing of guano. If you have not full faith in what we are telling you, try an experiment for yourself. Mix 200 or 300 lbs. of guano with two or three bushels of plaster, and that with two or three loads of charcoal dust from the bottom of some coal pit, or from burnt peat, or [Pg 49] swamp muck; or, if the charcoal is not attainable, use woods mold, or powdered clay or fine loam, to any extent you can afford; and if you can afford nothing but the guano and plaster, don't fail to afford a dressing of that, because it will afford you a rich return. No other manure can be used upon pasture land, to produce the same effect. Cattle never reject the grass of guanoed land, as they do that lately manured.

On Flax.—Experiments in England have proved guano superior to any other substance ever applied to this crop. With the aid of this manure, farmers will never complain of flax exhausting the soil. With 300 lbs. per acre, successive large crops can be grown upon the same ground. It should be plowed in, but not so deeply as for some other crops, as it is not expected to benefit succeeding ones as much as the present. As soon as the "flax cotton" movement now progressing is fully understood, there will be immense fields of flax grown for that purpose, and the best and most economical fertilizing material, and for which there will be a large demand, will be Peruvian guano; for no good farmer will attempt to grow a crop without it. A top dressing of 25 or 30 bushels of ashes to the acre will be found beneficial; but farmers ought to try which is best, more guano and less or no ashes, or the reverse. We cannot advise rotation with this crop, where guano is used, because the ground becomes so clean and free from weeds, it is of great advantage, and so far as we are informed, continuous good crops result from the annual application of the same quantity of guano, year after year.

On Cabbages.—Field culture. After the ground is well prepared, lay it off in checks three to four feet square. With a spade, throw out a deep spit at each check and put in a spoonful of guano, or at the rate of 400 lbs. per acre, and cover with soil. Set the plants immediately and water if possible. After the first hoeing, throw a handful of ashes on each plant.

For Carrots, Beets and Parsnips, plow in 500 lbs. per acre, twelve to eighteen inches deep. Top dress with ashes, salt, and fine manure in compost, to assist the young plants; the long roots will find the guano and it will produce such a crop as you never saw before.

On Hops. — Make a mixture of three cwt. of guano, one of salt, one and a half of saltpetre, and one of gypsum, for each acre; sow broadcast and plow in about four inches deep, and you will find your manure well paid for, and no exhaustion of the soil, as is usually the case wherever this crop is cultivated, as it is a very gross feeder, and requires very rich land or great deal of manure; for which reason it is not as much cultivated as it will be as soon as the virtues of the above application become fully known. [Pg 50]

For Tobacco, guano has been found to possess superior qualities, particularly in obviating the difficulty heretofore experienced in getting plants sufficiently early. We have the testimony of several witnesses to prove that burning a seed bed is quite unnecessary, if guano at the rate of 400 to 600 lbs. to the acre be mixed with an equal amount of ashes, and plaster and well raked in previous to sowing. Of the effect upon the crop, we give the testimony of a Virginia planter.

"In the spring of 1850, I applied 200 lbs. to the acre, on eight acres of land, which had been manured three years before for tobacco, and the same quantity, on three acres which had never been manured, and was very poor. On the last I also turned in some half rotted straw, raked up in the barn yard, after all the farm yard manure had been hauled out. Between these two pieces of land, 19 acres were heavily manured. The whole 30 acres had been well broken with four horses, early in the winter. The last year was the worst I have ever known for tobacco. Nevertheless, the first eight acres produced a very fine crop — the last three acres brought much better tobacco than the adjoining manured land, I should say not less than 600 lbs. to the acre."

Wheat on Guanoed Tobacco Land. — This field was sown with wheat, and the writer says — "I measured from these 30 acres next year upwards of 600 bushels of wheat of very fine quality; both pieces of guanoed land being *above* the average of the whole lot. Adjoining the *three* acres is an equal quantity of land of the same quality, which did not yield five bushels to the acre."

Of the effect upon another crop of wheat, the same gentleman says—"Two years ago I purchased three tons, two of which I applied to 20 acres of a James River hill, which though not gullied, had been a good deal worn by hard croppings, or bad cultivation, or both combined. The Guano was sowed *dry*, and on the wide rows laid off for sowing wheat, and ploughed in with two horses, the wheat then harrowed in. I forgot to say that the land had been fallowed in with three horses in the month of August, and the wheat sowed in October. In consequence of the dryness of the guano, and the width of the rows, the wheat was very much striped, being very luxuriant where the guano fell in the largest quantities. The product did not exceed 200 bushels, or 10 bushels to the acre, but the quality was so superior that I saved it all for seed."

"The land sowed two years ago, is now *striped with clover*, as it was with wheat."

This land is a tenacious red clay formation, from which the soil we presume has all been washed away "long time ago." No planter, he says, would have put such land in tobacco without heavy manuring; and yet it produced a fair crop of tobacco. Owing to distance from na [Pg 51] vigation, he could not use lime, or any heavy manure, and without guano he could not make crops, and, consequently could not make manure at home.

The editor of the American Farmer, in a note says—"Our correspondent appears to desire that his land should be brought to a state of fertility by the *quickest* practicable process, and from the beautiful results of his experiments with guano, we know of no agent to which he could look with so much certainty of success as to that very manure."

The quantity per acre for Tobacco.—We should recommend at least 400 lbs. sown broadcast and plowed in, on such land as described, not over four inches deep. The tobacco to be followed with wheat, the wheat with clover, the clover after one year with corn and then tobacco and guano again. The clover should have a bushel of plaster fall and spring. Whoever tries this will find the benefit of guano on tobacco. But there is one still greater benefit; we have been assured that the tobacco worm which it was supposed from his natural taste, nothing could nauseate, actually gets sick of guano, and refuses his accustomed food.

Another mode of applying it to tobacco has been practised successfully as follows:—Mark off the land in checks and put a small spoonful in each check, and cover up directly under the bed where the plant is to stand, three or four inches deep. To this a handful of ashes and plaster may be advantageously added. Guano does not give tobacco the rank flavor that is often acquired from high manuring.

Mr. Pleasants, although many experiments have failed, principally, as he believes, from improper application, says in a recent letter—"There is no actual reason why guano should not act as well on tobacco as any other crop. The failures are doubtless to be ascribed to the injudicious manner in which it has been applied. I can conceive of only one mode in which it can be used to advantage, and that is by strewing it along a deep furrow as described for corn; then bedding upon it and confining the cultivation to one direction. This has been my way of cultivating cabbages for the market for several years, and the guano has always acted promptly and powerfully. If chopped in at the base of the hill it would require a great quantity of rain to dissolve it and make it available to the young plants, for the conical shape of the hill has a tendency to shed the rain instead of absorbing it. I expect soon to receive very accurate results of a crop grown with guano, which Judge Nash represented to me as splendid. If I cultivated tobacco, I should have every confidence of success by planting it on ridges with the Guano buried at a considerable depth, say from four to six inches beneath the surface of the ridge—1 lb. to ten yards would be a sufficient quantity.

"In short, I consider guano good for any crop. For potatoes (that is Irish potatoes) I regard it as a specific manure. The quantity I ap [Pg 52] ply is 3/4 lbs. to every ten yards put in the furrows as recommended for corn and tobacco, and then covered over about one inch with earth drawn from the sides of the furrows. After this the potato cuttings are planted and covered over with the plough or hoe. The quantity recommended is about right as far as my experience goes (which is of several years duration) if the cuttings are placed about two inches apart."

Guano for Cotton. — But few trials upon this crop have come to our knowledge, but such as have, indicate that it will prove one of the most valuable promoters of the growth of this staple product of America ever discovered. The analysis of cotton — stalk, seed and lint — compared with that of guano, is sufficient to prove the latter to be the very matter required to produce the former. We are assured upon the most reliable authority that guano will give an average increase of pound for pound upon any soil producing less than a bale per acre so that every pound of guano costing two and a half cents, will give a pound of cotton averaging at least 6¼ cents.

Mode of applying on Cotton Land. — Open a deep furrow and drill in the bottom at the rate of 400 lbs. to the acre, upon land usually producing 300 to 500 lbs. seed cotton, and less for a better quality of land, down to one-fourth the quantity. Bed on this as deep as you please; the moisture of the earth will disengage the ammonia and phosphates, and send their fertilizing properties up to the roots. Never use guano as a top-dressing for cotton. The seed will be found better matured, and consequently more valuable to manure another crop, besides being so much easier separated from the lint, which will be found as much improved in quality as quantity. For Sea Island planters, where manure is so valuable and so hard to obtain, we would earnestly recommend a thorough trial of guano. As the land for this crop is mostly prepared with hoes, care must be taken that the servants do not neglect to bury it at the very bottom of a good bed.

From the knowledge the writer has of the culture and value of long staple cotton, and the price and value of guano, he has no hesitation in expressing his honest conviction that a clear profit of two to four hundred per cent. may be made upon every dollar expended in the purchase and proper application of guano to that crop.

Guano, for all staple crops in the United States, is no longer an experiment. It has been clearly demonstrated, to be the cheapest and most valuable fertilizer, particularly for all poor, worn out, hard used and exhausted soils ever discovered; which no sensible man will neglect to profit by, as soon as he learns its value, unless prevented by deep prejudice or strong circumstances.

Application to Miscellaneous Crops.—Under this head we will give the experience of several individuals in various sections, soils and climates, [Pg 53] in hopes it may encourage the doubtful, and direct those who are disposed to emerge from darkness into the light of scientific agriculture. A gentleman from Warsaw, Virginia, where the soil is generally a sandy loam, badly worn by long years of bad tillage, says, "My wheat looks finely, especially where I applied guano last fall. I put it in with the seed furrow about three inches deep, and also with double plow six inches deep, harrowing in the wheat frequently side by side. At this time I can see no difference in the wheat crop. I use a large wooden toothed harrow extending over the bed of ten feet, and an even soil, free from stone; they do admirable work and drill the wheat as if put in with the drill."

Willoughby Newton, whose operation we have already spoken of, says; "I do not believe it possible to improve a farm, on the old three shift system, of corn, wheat and pasture, without a large supply of foreign manures. If clover can be substituted for pasture in the summer, then the land, if not naturally poor, may be rapidly improved by the use of lime alone, in addition to the putrescent manures that may, by proper care, be made on the farm. On other land of less fertility, and drier, I greatly prefer the five field system, under which, with the use of lime, guano and clover, a rapid improvement may be effected at the same time that heavy crops of wheat are reaped."

Another writer in speaking of how to improve worn out lands, says; "Let whatever little surplus he can spare from supplying the necessary wants of his family be laid out in the purchase of some one of the reliable concentrated manures. [Guano is by far the cheapest, and therefore the best for him, if he will plow it in well]. And my observation and experience have convinced me that he may make such improvement as will bring him a quick return, and soon enable him to get his farm well set in grass. This once effected, his facilities for its further improvement will assuredly increase in a ratio just in proportion as he is careful to pursue the course indicated. If a farmer can succeed in getting his fields well set in grass, a large and long array of facts and experience have proved that he may then, under a judicious course of management, render them more and more fertile without foreign aid of any kind whatever."

The editor of the American Farmer, in deprecating the price of guano says, "Of the efficacy of guano, in restoring worn out lands to productiveness—of its capacity to increase the yield of any lands in a sound condition—there cannot be a doubt; but even with all its regenerating properties, we do think that its market value is too high. Forty-eight dollars for a ton of 2,000 lbs. of Peruvian guano is more than it is intrinsically worth, and should it be continued thus high, must, we should [Pg 54] think, limit its use, for the obvious reason, that farmers cannot afford to pay a price for it which is so disproportionate to its real value."

Yet they do continue to pay, and make it pay a greater profit than any other manure ever purchased. We hold to have done as much as any other individual to reduce the price of guano, and wish as heartily as does the editor of the Am. Farmer, it was only half the price it now is; yet, we must counsel our readers not to wait for that cheap time coming. It is now cheaper than it was then, and probably as low as it will be for years; and in the hands of the present agents, the public may depend upon a regular supply, and of genuine quality, at what the Peruvian government deem a fair price.

Guano for Melons and other Vines.—Mr. Pleasants, of whom we have before spoken, and whose long experience in the use of guano in connection with a market garden, entitle him to a high degree of credit, says, "I have been in the habit of using it for several years, and can testify to its value, not only using it for melons, but for the whole tribe of cucurbitacæ. The mode of application which I prefer is this; when the ground is prepared and checked off, remove the loose soil at the intersections of the furrows, leaving clear spaces on the substratum of not less than eighteen inches in diameter. Upon these spaces sprinkle guano, at the rate one pound to eight hills. Follow with a hilling or grubbing hoe, and incorporate the guano with the subsoil; then draw the loose earth back, and finish by chopping a small quantity, a spadeful or less, of well rotted manure into the hill near the surface. Guano placed near the surface, will remain almost inert, and buried deep, as I recommended, it will be too remote from the seed to give the young plants the quick start which is indispensable to an early crop of melons. The small quantity of manure near the top of the hill answers the purpose of immediate forcing, and enables the roots to strike rapidly into the guano, when the growth of the vines will be stimulated to such a degree as to cause them to mature their fruit a week or ten days earlier than they would do from either guano or manure alone. Melons equally fine may be raised from nothing but guano, applied in the manner directed; but they will not be an early crop, from the fact that the plants remain almost stationary until the roots reach the guano. Last year, from such a preparation as is now recommended, I had as fine a crop of melons as I ever saw; and they began to ripen at a very early period in the season. Two years ago, I had them nearly or quite as good from guano alone; but they were late. This year the crop was almost a failure, from the wetness of the season, which caused the vines to die. Cantelope melons, however, have produced abundantly, grown entirely with the aid of guano. Where manure is scarce, I have no doubt an admirable compost might be prepared, consisting of guano and rich [Pg 55] earth. It should be made several weeks, or even months, before it is wanted for use; and the heap worked over frequently in order to bring it into a suitable condition. Such a compost would doubtless supply the place in the hill which I have assigned to the manure. For pumpkins, squashes, cymblins and cucumbers, when it is not particularly desirable to have them early, nothing more is necessary than to prepare the hills with guano."

The following extract from a letter of E. G. Booth, to F. C. Stainbrook, written in that plain familiar style of one friend to another, which characterises the man, with an evident intent to do good; though it was not designed for publication, we give it because we believe it will do others good, as well as the recipient. Mr. Booth confirms our opinion often expressed, that the poor old barren fields of lower Virginia, are really more valuable than the rich lands of the west; because, owing to facilities of intercourse with commercial cities by water, these lands can be bought, and cultivated by aid of guano, with more profit than the richest prairie farm in Illinois. Mr. Booth's testimony upon the durability of this manure, is enough to contradict all the assertions that "it is of no use for only one crop."

On his land, strangers can easily tell where guano was applied four years previous.

"Yours of the third has been received, and it affords me pleasure to give you any information in my power. The wheat crop during the the winter was very unpromising. There was a general complaint that it was too thin. The Poland wheat (most generally sown in this neighborhood,) is said to branch more than other kinds, and I regard the present prospect of the wheat crop as flattering, particularly where guano was used. It is now a fixed fact, that no poor land ought to be cultivated without guano, by any person who can command the money or credit to buy it. It is remarkable that it pays a much better profit, or per cent. on the investment, on poor land, than rich. I was inclined for some time to believe that the difference was really in appearance alone. The difference of five bushels increase on land which without it would bring only fifteen — or in other words, an increase from fifteen to twenty bushels to the acre, would not be very perceptible, while an increase of five bushels on land previously making only five, would be very evident. Still, the real increase would be five bushels in each case. I am now however, decidedly of the opinion that it pays a much larger per cent. on poor than rich land; because it supplies that in which poor land is deficient, and of which rich land may have enough. I have it now in strips on a clover fallow, scarcely showing any difference. I last applied it on about the poorest land on my plantation, and the product was remarkable. This circumstance much reduces the difference between the value of poor and rich land, and admonishes us that there [Pg 56] is not a plot in our wide extended surface, which need be abandoned or neglected. We can, if we manage properly, support a population which will out vote the West in 1865. There is another fact which experience confirms, that is it is much more durable than at first supposed. My visitors have been able to point out the strips of land on which it was sown, four years after its application. I noticed a very evident effect on the farm of Mr. William Fitzgerald, a few days ago. He last year put it in drills, and hilled on them for tobacco, in the fall the whole surface was sown in wheat, which is now growing in ridges corresponding with the furrows where it was placed.

"While on the subject I will mention another fact different from first impressions, viz: that it is more productive, (the first crop, at least,) when harrowed in with the grain, on the surface, than when turned in very deep. I have yet to satisfy myself which is most durable. In the experiment which lasted four years, I think it was turned in. The purchases the ensuing fall will be very large. Those who were most incredulous are now going in largely. A very intelligent and enterprising friend of mine, who has been improving his land judiciously and profitably in this way, related to me an anecdote which occurred to him. He had two neighbors remarkable for their judgment and success in farming as well as other things, who, however, were inclined to underrate his expenditure of money in these elements of improvement. They knew he had purchased and used a ton of guano, and thought they knew where he had used the whole of it. They went, not exactly by night, but rather privately, to examine into the result. They made their observations and calculations, and agreed that he had got his money back, but no profit worthy consideration, and were only confirmed in their opposition to such an expenditure. The truth was, however, that only about one eighth of the ton had been used where they calculated for the whole. One of these gentlemen, I am informed, is now about the largest purchaser of such articles in the county; and perhaps the other also, though I have not been informed."

PLASTER WITH GUANO.

A Virginia farmer, in a letter of December 1847, in speaking of using plaster with guano, and the effect says—"I am a firm believer in the merits of the mixture, and always use it. I have used it on turnips with decided effect, as decided as that following any application of guano I ever saw. Several farmers of my acquaintance used the mixture of guano and plaster, and stable manure and plaster habitually, like myself, and one told me he used it half and half, producing the most marked effect on wheat, and that a neighbor of his had used it in the same pro [Pg 57] portion with the same effect—the usual surprising effect of guano. For myself, I used some $400 worth of guano on wheat this fall, the whole of it mixed with plaster. I believe the effect of the mixture will not be so vigorous on the first crop, as guano by itself—the plaster husbanding the ammonia for succeeding crops, upon which the mixture, (if the theory be correct,) will have more effect than guano unmixed, that being exhausted by the first crop."

A gentleman after making sundry careful experiments with plaster and carbonate of ammonia, thus expresses his conclusions—"These experiments prove to me that no matter in what state, (whether *wet*, *moist*, or *dry*,) plaster is presented to guano, or any other manure from which the carbonate of ammonia is escaping, it must retain a certain amount of ammonia that would otherwise be lost in the atmosphere."

The editor of the American Farmer says—"If the soil be poor, and it be desired to permanently improve it, at least four hundred pounds of guano, without respect to the fixer used, should be spread *broadcast*, on every acre of it, and plowed in to the full depth of the furrow. If the land be in moderate heart, three hundred pounds will be enough per acre. Where the soil may be good, two hundred will be sufficient. These quantities, as the reader will observe, have relation to broadcast applications, as all should be where general improvement is contemplated; if compelled to confine his experiments on corn to applications in the hill, a form of manuring, we have ever disapproved, two hundred pounds, or even one hundred of guano, will manure an acre, mixed with a bushel of plaster, five bushels of slaked ashes, and a double horse cart of wood mould more effective than ten loads of manure applied in the hill."

Yes, as has been proved by careful experiment made in England, more than fourteen tons of manure. The editor also says, what we have so often repeated—"We hold these to be agricultural truths—that guano is most beneficially applied, when ploughed in as spread on the the earth, never less than four inches deep—and better, for permanent effect, to be ploughed in deeper, say six to eight inches—where it may be desirable only to bury it four inches deep, the land should be previously ploughed as deep as the furrow can be turned up, and the guano applied at a second ploughing—that all top-dressings with guano are wasteful, inasmuch, as from the volatile nature of the more active parts of the manure, great loss must inevitably result from all such applications, and because, more moisture than is to be found on the surface, is necessary to excite, and carry on, that healthful progressive state of decomposition, which is required to render guano most available for present production and future improvement. [Pg 58]

"We do not hesitate to express the opinion, that when properly used, as an adjunct to lime or marl, that it will bring up any sound worn out land, to at least its original degree, if not a greater degree of fertility; provided its application be followed by clover. We believe that, when properly applied to land, either limed or marled the previous year, it will add twenty-five, thirty, and, in some instances, forty per cent. to the product of wheat; besides infusing into the soil, the capacity to grow luxuriant crops of clover, and thus fit it for profitable future culture. If it will do this, and we are certain it will, then it will achieve all that any agriculturist can reasonably expect of it, or of any other fertilizing agent; and we are very sure there is no other manure equally efficacious, within the reach of farmers and planters.

"Guano differs much in quality; that from Peru, is confessedly best of any which has yet been submitted to actual experiment by agriculturists, or tested by the analysis of chemists, being much richer in its nitrogenous element, than either the Patagonian or African variety."

He also says—"400 lbs. of guano and 1 bushel of plaster, will ensure a good crop of corn, so will 200 lbs. guano and eight bushels of bone earth, or 20 bushels of bone earth, 10 bushels of ashes and 1 bushel of plaster. Each to be ploughed in."

Much more might be said in favor of using plaster with guano, or some other fixer of ammonia, wherever it is exposed, on or near the surface. We add a few more extracts mainly to show that deep ploughing, and plentiful manuring, are the sure guarantee of bountiful crops. Bone-dust, except when used in the drill, should always be harrowed in. It should be put in bulk with other matters, and excited into an incipient state of decomposition before being used.

Guano should always be ploughed in, if practicable. Harrowing and cultivating in guano "have been practised both in this country and in England, by intelligent farmers; and in various instances have been spoken approvingly of, success having attended such applications in single crops; but we doubt whether much, if any permanent benefit were done to the soil, in qualifying it for the production of the subsequent crops of a course of rotation. In Peru it is used topically, but such applications are always followed by immediate irrigations of the soils to which it is applied, the Peruvians acting upon the philosophical principal, whether they comprehend its theory or not, that to secure the nutrient properties of this active fertilizer to their growing crops, it is essential that they provide an absorbent, and that they find in the water furnished by their processes of irrigation. Experience, practice, and irrigation have taught them, that unless they cause the carbonate of ammonia, and the various compound substances with which it exists in the guano, to descend speedily to the roots of their plants, that from the vo [Pg 59] latility of its more active and efficient elements, they will be expelled by the heat of the sun, escape into the air, and be lost for all the purposes of vegetable growth.

"But in view of the whole ground, taking into consideration the evanescent nature of any ammonia in guano in the compounds in which it exists, to be converted into that form, we honestly believe, that so far as lasting benefit to the land may be concerned, guano should be ploughed in.

"In all tolerably good Guano, there is a sufficiency of the carbonate already formed to carry on healthful vegetation, and therefore, it is best to place it sufficiently deep to prevent the waste of an element so essential to the growth of plants, and so liable to loss.

"It is possible where the soil had been, by repeated harrowings, reduced to a state of very fine tilth, that guano may be covered sufficiently deep with the Cultivator to become mixed with, and consequently be absorbed by the vegetable remains of the earth, and thus be prevented from loss by escape of its volatile gases; especially would this be the case, if the process of cultivating it in, were soon after followed by penetrating rains. In admitting this, we still adhere to the opinion, that so far as permanent benefits are concerned, the most economical mode of applying guano to the earth, is by the plough.

"As soon as the guano is ploughed in, the wheat should be sowed and harrowed in, in the usual way. In our climate we can sow wheat on the poorest corn ground late in November and have as fine a crop, and harvest it as soon, as we can obtain from well prepared and fallowed without guano sowed early in the season, For every 100 lbs. of guano, not exceeding 250 lbs. we calculate on reaping of an average season from six to seven bushels, sometimes eight. From a greater quantity though the product will be increased, yet it will not be increased in the same proportion, and 200 lbs will also be sufficient for the production of two good grass crops following the wheat and will then leave the land in an improved condition."

Charcoal and Guano.—The benefit of charcoal with guano will be understood from the following extract from "Scientific Agriculture," on the nature of charcoal and its use as a manure.—"Charcoal on account of its power of absorbing gases and destroying offensive odors, is a valuable addition to the soil; its operation as a manure is not so direct as some other manures; that is, it is not so useful on account of any element it furnishes to plants, as by the intermediate office which it performs, of absorbing and retaining in the soil those volatile matters which plants require, and which would otherwise escape and be lost. It is beneficial as a top-dressing, and as an ingredient in composts; it evolves carbonic acid in its decomposition, and is in this way directly [Pg 60] useful to plants. Its powerful antiseptic properties render it very useful to young and tender plants, by keeping the soil free of putrifying substances, which would otherwise destroy their spongioles and prevent their growth."

And its capacity to absorb many times its bulk of gaseous matter, will always give it value as an absorbent of escaping ammonia from surface dressings of guano.

The editor of the Farmer also says—"In our climate, we should be opposed to all topical applications of any strongly concentrated manure like guano by itself,—and, indeed we should, under all circumstances, prefer to have it ploughed in, if practicable; but as we presume our correspondent has been prevented by circumstances, from using guano at the time of ploughing for wheat; and of course, must avail himself of the next best plan of deriving benefit from its use, we would advise, him next spring, as soon as the frost is out of the ground, and it is in a state to bear a team; to mix, in the proportion of 100 lbs. of guano, one bushel of fine charcoal, and one peck of plaster per acre, then to sow the mixture over his wheat field, lightly harrow the ground, and finish by rolling; and we have no hesitation in saying, that his wheat crop will be benefitted more than twice the cost of the manure. We say to him farther that he need not fear injuring his wheat plants by the operation of harrowing and rolling; for, on the contrary, it will act as a working, and prove of decided advantage. We feel very certain that the admixture of charcoal and plaster with guano, together with the covering it will receive by the harrowing, will prevent any material loss of the ammoniacal principles of the latter; as independent of the affinity existing between charcoal, plaster, and all nitrogeneous bodies, they will be greatly aided by the vital principle of the plants themselves. We are not, however, left to the lights of theory alone, in this matter, but have the experience of the Honorable Mr. Pearce, of Kent county, of this State, to guide us to a practical result,—he used, some years since, a top-dressing of guano and plaster upon his wheat field, and was rewarded by a large increase of crop."

A correspondent says—"I am satisfied from experience and observation in the use of guano for the last twelve years, that the best method, decidedly, of applying it to our crops in this dry climate, is to plow and spade it into the ground; and autumn is the best season for doing this, as it gives time for the pungent salts contained in the guano to get thoroughly mixed with the soil before spring planting. Do not fear to lose the guano, by plowing it as deep as you please — it will not run away, depend upon it. At the south it loses half its virtue if not plowed in at least three inches deep; six to twelve inches would be still better. [Pg 61]

"Spread broadcast on grass land, late in the fall or early in the spring, if not plowed in before sowing buckwheat, rye or wheat, then spread it broadcast after sowing the grain, and harrow well and roll the land. This last operation is quite important."

Value of Guano on account of its Phosphates. — He who wishes to have the best grazing grounds, where he can present the richest and most nutritious herbage to his cattle, will keep his ground well supplied or manured with guano that abounds in phosphates, knowing that it will supply the needed nutriment to the grass, and by the grass to the cattle; and thus his stock will be kept in a high condition and full flesh, either for the farm or the market.

Again; he who raises wheat, corn, or other grains, has an equal inducement to look to it that his manures are abundantly impregnated with these essential elements. Phosphates, so available to the raiser of stock, are equally so to the producer of grain; because the size, richness, and nutritious qualities of the grain depend largely on the presence of these in the soil. A farmer, therefore, has a vital interest in this matter, and should obtain what best suits his purpose. The most intelligent English farmers are so well convinced on this point, that substances containing only ten per cent. of phosphate of lime, are sought after, dissolved in sulphuric acid and water, and sprinkled on the soil. Bone dust also is used, and to a certain extent, is available, because one of the principal constituents of bones, is phosphate of lime. But the article in which the phosphates are the most convenient, because the most minutely distributed, is guano; and this, when judiciously used, must find favor wherever it can be obtained.

That which contains a large proportion of phosphates, in combination with ammonia, nitrogen and alkaline salts, apparently in the exact proportion required by nature, such as analysis and experience proves is the case with Peruvian guano, will be sought after by every farmer who reads the evidence of its value which we have given in these pages.

It is idle to talk of bones to restore the waste of phosphates in the soil that is being constantly carried away in grass and grain, beef, pork, mutton, milk and cheese, much of which passes into the sea from the sewers of cities, to be there retained in that great reservoir for the future use of men. It is from that we are now drawing our present supplies. Happily for mankind in all civilized countries, the discovery of guano has, in a providential manner, met the very wants of the times, in reference to the reinvigoration of certain kinds of soil, since this manure furnishes the elements most needed to supply the waste arising from cultivation, and to develop vegetation.

The impossibility of procuring bones enough to supply the wants of the comparative few now engaged in using guano, may be readily learned [Pg 62] by any farmer who uses ten tons of guano per annum, if he will undertake to "pick up bones" enough to furnish him the same amount of phosphates contained in that quantity of guano. Then if all who are now using it, would drop guano and take to bones, it would soon be found to be hard picking. Save all the bones and apply them to the soil, is a standing text with us; upon the same soil use all the guano your can procure and you will not need to pick bones—you will grow bones to pick. It may be very patriotic to talk about expending the money at home, for bones, instead of sending it to Peru, for guano; but that talk is all for Buncombe, there is not a particle of sound reason in it. If all the bones in the United States could be saved and applied to the land again, we should still fall short of a supply, and be obliged to do as England did before the introduction of guano; go about and ransack grave yards of great battlefields, for more bones. With all the guano imported, or that will be imported, and all the bones that will be saved, there will still be room for more phosphates in the millions of acres of hungry soil in America. What would be the effect if a few such farms as Willoughby Newton's, and Col. Carter's, who each use 30 to 40 tons per annum of guano, should come all at once into the bone market for their supplies. In our opinion there would be such a rattling among the dry bones, we should hear no more about substituting them for guano. The fact is an incontrovertible one, that nothing on earth nor under the earth, or in the sea, has ever been discovered, which can be used as a substitute for guano. Its small bulk is alone sufficient to commend it to favor.

The Royal Agricultural Society of England offers a prize of £1,000 and the gold medal of the society, for the discovery of a manure with equal fertilizing properties to the guano, of which an unlimited supply can be furnished in England, at £5 per ton.

"*Analogy between Bones and Guano.* — There is a striking analogy in composition between bones and guano, which is, for other reasons interesting to the practical man.

The following table exhibits the composition of bones compared with guano, supposing both in the dry state. Bones, as they are applied to to the land contain about 18 per cent. of water. Ichaboe guano from 20 to 25 per cent.

	Bones.	*Guano.*
Organic animal matter,	33	56
Phosphates of lime and magnesia,	59	26
Carbonate of lime,	4	6
Salts of soda,	4	10
Salts of potash,	trace	trace
Silicious matter	0	2
	100	100"

And these substances are found in guano already in a pulverulent state, while bones have to be reduced by mechanical or chemical means to the same condition before they are of any use as manure. Do not, we again repeat most emphatically, do not waste a bone; dissolve all you can get in sulphuric acid and mix with guano — save and make all the manure possible, both by the stable, compost heap and green crops, and then you will have money to buy guano, by which you can save the immense labor of hauling to distant fields, and still have the satisfaction of seeing them as fertile as those highly manured near home.

When the farmer raises crops for sale, and removes his grain and grasses from the land, he sells a portion of his soil; and if he does not renew in some way the saline matters taken away in his crops, he invariably impoverishes his farm. This work of exhaustion is now going on to an alarming extent, and the prolific wheat lands are to be searched for farther and farther westward as the operation proceeds.

Every one knows the superiority of wheat grown on newly cultivated lands, and most farmers are aware of the fact that soils become exhausted of something, they know not what, but of something essential to the most favorable production of grain. This something is found in guano, and by it the original fertility of land can be more easily, more certainly and cheaply restored than by any other means as yet discovered.

Professor Mapes in one of his letters of advice says; "As no farm, under ordinary usage, will supply as much manure as may be used upon it with profit, I am glad you intend to use guano, as it is an admirable manure, replete with many requirements of plants. The ammonia of the guano is in the form of a carbonate, and therefore so volatile as to escape from the soil into the atmosphere before plants can use it.

"You will readily perceive, therefore, that the sulphuric and phosphoric acids require amendments, and the ammonia should be changed from a carbonate to a sulphate of ammonia, which is not volatile. All this may be readily done by dissolving bone dust in dilute sulphuric acid, mixing it with the guano, and then with a sufficient amount of charcoal dust to render the mass dry and pulverulent. The more charcoal dust the better, as it absorbs and retains ammonia, and after it is in the soil, will continue to perform similar offices for many years, only yielding up ammonia as required by plants, and receiving new portions from rains, dews, &c."

If used as a top dressing, this change from a carbonate to a sulphate may be necessary; but not so if well mixed with the soil, particularly one in which clay predominates. In such a soil it is not even necessary to adhere to the direction to plow the guano deeply under. If it is but slightly harrowed in, the nature of the clay is such it will prevent the [Pg 64] escape of the ammonia. If you require phosphates, more than ammonia, add the superphosphate of lime; but in no case omit the guano.

Use of Salt with Guano.—Common salt at the rate of a bushel to 100 lbs. of guano, well mixed, may be used to good advantage either as a top dressing, or when plowed in. The effect of the muriatic acid of the salt upon the guano will be, as both are dissolved in the earth, or by dews and rains, to form muriate of ammonia, which is not volatile; consequently the salt prevents loss by exhaustion, which is sure to take place when the guano is used as a top dressing, unless prevented by something to act as a fixer of the ammonia.

The wisdom of this law of nature in making the most precious saline manure a fixed and difficultly soluble salt, is at once obvious; for it is thus kept always ready in the soil for the plants to act upon according to their need. If we cut plants down before the seeds form, we have all the phosphates the plants contain diffused throughout them, and if we allow the seed to ripen, the phosphates, as before observed, will be found mostly in the seed. We find them in the state of phosphate of potash, phosphate of soda, phosphate of magnesia, and phosphate of lime, and probably, also, phosphate of ammonia. Now all these salts are essential to the growth and sustenance of animals, and without them grain would cease to be sufficient.

The necessity of restoring inorganic substances to the soil, may be better understood by examining the following table:

Mr. Prixdeaux states that the following quantities (of inorganic matters) are removed from an acre of soil by a crop of wheat, of 25 bushels of grain, and 3000 lbs. of straw—

	By the grain.	*By the straw.*	Total.
	lbs.	lbs.	lbs.
Potash,	7.15	22.44	29.59

Soda,	2.73	0.29	3.02
Magnesia,	3.63	6.99	10.62
Phosphoric acid,	15.02	5.54	20.56
Sulphuric acid,	0.07	10.49	10.56
Chlorine	0.00	1.98	1.98
	— —	— —	— —
	28.60	47.73	
Gross weight to be returned to an acre,			76.33

Professor Johnson says—"Soils are barren either from the presence of a noxious principle or the absence of a necessary element. It is therefore highly important to be able to distinguish between the two cases. [Pg 65]

"The art of culture is almost entirely a chemical art. Its processes are explained on chemical principals in part, but partly on mechanical and natural ones.

"All forms of matter may be divided into one of the two great groups—organic or inorganic matter."

In Peruvian guano, both these substances exist in a better and cheaper form than can be obtained from any other source.

The editor of the Genesee farmer, whose scientific information none can dispute, strongly corroborates this opinion. In a late number he says—If we admit that phosphate of lime is a necessary ingredient in a special manure for wheat—Peruvian guano would at present be much the cheapest source of it; for, in addition to the 16 per cent. of ammonia, it contains 20 per cent. of phosphate of lime in first-rate condition for assimilation by the plant, as well as other fertilizing ingredients of minor importance.

As a manure for wheat, therefore, we greatly prefer good Peruvian guano, even to the *improved* superphosphate of lime.

Difference in favor of Guano over Bone dust. — Robert Monteith, England, dressed oat ground with 276 lbs. guano per acre, cost 31 shillings, produce 59 bushels, value £7 7s 6d. Same quality of land with 10 bushels bone dust, cost 23 shillings and fourpence, produced 43 bushels value £5 7s 6d, which gives a balance in favor of guano of £1 12s 4d, or about $7 50 per acre.

Difference in favor of Guano over Manure. — The Yorkshire Agricultural Society of England, instituted a series of experiments several years ago for the purpose of working out practical facts in relation to guano, through a series of crops, upon different soils, by different persons, upon whose report the utmost reliance might be placed, so as to determine the value, or advantage to British farmers, who might use this extraordinary fertilizer. This report has just been published, and the following is a synopsis of the results. The experiments were arranged under the following heads —

1. To show the natural produce of the land, one part was to have no manure whatever.

2. Was to have twelve tons per acre of farm-yard dung.

3. Was to have six tons of dung, and one cwt. each of guano and dissolved coprolites; and

4. Was to have two cwt. of guano and two cwt. of the coprolites.

Other substances might be tried as additions, but these were to be the standard experiments.

Mr. Cholmeley's turnips, grown on a loamy soil had the heaviest crop on No. 3, the dung, coprolite, and guano, beating the farm-yard manure by some 5¾ tons per acre. [Pg 66]

Mr. Johnson's experiments were tried with various manures singly; and his Peruvian guano gave the greatest weight of the class of substances tried; but 10 cubic yards of farm-yard manure had previously been applied to the whole land.

Mr. Maulevere's heaviest weight, also applied singly, was with the 12 tons of dung; but only 14 cwt. more than the dressing with 2 cwt. of coprolites. This soil was a light clay.

Mr. Newham's on a limestone soil, were the heaviest with No. 3 — the same as Mr. Cholmeley's — and were 16 cwt. heavier than an application of dung alone.

Mr. Outhwaite's, on a hungry gravel, were the heaviest, with 9¾ tons of dung and 2 cwt. of guano, for all the land had been dunged at this rate, and exceeded 14½ tons of dung by 2 tons 9 cwt. per acre.

Mr. Scott's were the heaviest on No. 4, — the guano and coprolites, and 1 ton 7 cwt. more than 20 tons of dung, — his soil was a strong loam.

Mr. Wailes's were the heaviest, with 4 cwt. of coprolites, showing an increase over 20 tons of dung of 2 tons 9 cwt. per acre; the soil is a useful loam.

The first fact which strikes the observer, is, that as a general rule, there is not only an addition to the crop by the addition of those artificial manures, but there is, in some cases, more absolute crop produced by them than by farm-yard manure alone.

Now to bring this to the test of figures, the coprolites at £5 per ton, and the guano at £10 per ton, will be at the rate of 2 cwt of each, £1 10s per acre. Now assuming this to be equal to 20 tons of dung per acre, we should require to be able to produce the dung at 1s 6d per ton to cost us the same money. But it can be neither produced nor purchased at any such money. In the whole of the cases referred to, the manure is most costly, and yet we find hardly any case where there is not an addition to the crop, of say two to three tons of turnips per acre, by such an increase of manure as the guano. Now, if a ton of turnips be worth 10s., or even 9s, there is at once an element of repayment; for, if a soil be in a condition to give a large crop of turnips, it is almost certain to be capable of giving a large crop of any other plant to succeed.

Mr. Charnock gives it as the result of his practical experience, that 4 cwt. of Peruvian guano, without manure, is the cheapest and best mode of growing turnips; but the general testimony seems to be decidedly in favor of what all farmers find it the best and easiest to do, viz., to add a small quantity of artificial manure to that which the farm will supply, and so to spread the whole over the land, rather than put all the dung in one place, and all the artificial manure in another.

No one can doubt the true statement of this report, which proves $7 50. worth of guano equal to 20 tons of manure—reducing the worth [Pg 67] of that to one shilling and sixpence—about 34 cents—per ton, or one dollar a cord. Now, as manure is often estimated in this country by the cord, and valued at about $4, and applied at the rate of 6 cords per acre, it follows that a saving of $14 50 per acre may be made by using 250 lbs. of guano instead of purchasing the manure. This Yorkshire experiment exactly corresponds with those made in this country, some of which we have detailed, and which proves that a farmer cannot buy manure at the common selling prices; and if he hauls his own the distance of a mile, he will expend more value of time, than it is worth to him on the land; because the same value of time—"time is money"—expended for guano, will bring him better returns. In this, as before stated, we are confirmed by Professor Mapes; and here is the opinion of Mr. Hovey of Boston, the eminent horticulturist, which we find in the August No. of his magazine, as follows—

"If, after such evidence as this, farmers will continue to buy ashes at eight cents a bushel, or manure at three to six dollars a cord, including carting, and use them alone, then let them do so, but they should not complain that their crop cost more than it comes to. To orchardists and fruit growers, this information is of the greatest value, and we trust they will not let it pass unheeded."

This opinion is valuable because it has been stoutly asserted, that however well guano might answer at the South, it was of no use in the hard soil and cold climate of New England. This is a fallacy which will soon be cured by knowledge, and self-interest is a very strong prompter towards the acquisition of the knowledge, that guano is the best, cheapest, most suitable, convenient and productive manure ever used by a New England farmer, and just as suitable for that climate and soil as it is for Virginia. We assert, without fear of successful contradiction, that there is not a farm—not a field—covered with five-finger vines and mullens, in the State of Massachusetts, which may not be made to produce as profitable crops, by the use of guano, as any Connecticut river farm. Farmers are about the hardest class of men in the world to learn new doctrines; or that science has anything to do with the business of this life, and what all other life in a civilized country is dependent upon. Yet science teaches, by unerring truths, that the plants the farmer cultivates, are composed of carbon, obtained by plants chiefly from the soil and atmosphere; oxygen and hydrogen, obtained by plants chiefly from water, carbonic acid, &c.; nitrogen obtained by plants chiefly from manure, and also from rain and snow; silicium, in combination with oxygen, called *silicia* or sand; lime in combination with phosphoric and other acids; potash and soda in combination with acids; magnesia, in combination with acids, and various oxides of metals, the presence of which, however, is not very important, as they ex [Pg 68] ist in an exceedingly small quantity. And that guano is composed of ammonia (formed of nitrogen and hydrogen,) combined with carbonic, oxalic, phosphoric, and other acids; lime, combined with phosphoric oxalic, and other acids; potash and soda, combined with muriatic and sulphuric acids; magnesia, combined with phosphoric and other acids; animal organic matter, containing carbon, and also nitrogen.

Now, is it not enough to prove that all the ingredients, with the exception of the metallic oxides, exist in guano, which are required by the plants grown for the sustenance of man.

Putting guano into the soil, therefore, as a manure, is clearly restoring to the earth those substances which plants abstract from it, and which are absolutely necessary for their growth.

The questions, then, which the farmer should now ask are, "which is best for me to buy, guano or coarse manure?" The evidence just given answers that question. "I have manure, teams, and men to haul it; my fields are from one to three miles distant, is it economy for me to let my teams lay idle and buy guano?" By no means. But you can probably employ men and teams in other improvements to much better advantage. With your manure make all your home lots exceedingly rich. With your men and teams clear off stones, dig ditches to put them into, drain your land, or build fence—bring bog meadows and swamps into dry cultivation—send every little brook through artificial channels for irrigation—send water up from lowland springs and streams by hydraulic rams for the same purpose, and for stock on the hills; or bring it down from hillsides if you are so situated; and buy guano for those distant fields, instead of wasting time in the laborious operation of hauling manure. Those who use guano, are enabled by the saving of time, to say nothing of their increased profits, to make improvements which are utterly impossible to accomplish under the old system.

How to choose Guano. — As we are satisfied no sensible reader can have perused the preceding pages, without having come to the determination to make a trial for himself, we will give him some general instructions about buying guano.

In the first place, we lay it down as an incontrovertible axiom, that the Peruvian guano, at the current price for years of that and all other, is the cheapest and best, because it contains the largest amount of ammonia, in a perfectly dry state; as a carbonate, true, but because dry, it is permanent and not likely to loose by volatilization by long keeping.

If other varieties contain a larger proportion of phosphates, and are sold at a less price, experience proves they are not cheaper. If an additional quantity of phosphates is desirable, it can be obtained in a cheaper form from dissolved bones, or bone dust and shavings of bone workers; or from mineral phosphates of lime. Recollect, guano under [Pg 69] no other name, has ever equalled the Peruvian, in the results as compared with the quality or cost.

Therefore buy none but Peruvian. To guard against deception, be careful of whom you buy. If you cannot buy directly from the agents, be sure the character of your merchant is a sufficient guarantee against adulteration.

To test the quality of Guano. — The best test is the price. Unlike other merchandise, this article is not subject to fluctuations. Being a government monopoly, the price at which the agents are to sell here is fixed in Peru, and that price may be easily known; therefore, if any dealer offers you Peruvian guano at "a reduced price," you may be sure the quality is reduced also. Remember, that the lowest price by the ship load, it can be procured for of the agents in Baltimore or New York is $46 per ton of 2240 lbs. To this, every fair, honest dealer, must add freight, insurance and profit. Every man who sells without such addition, you may be sure will make his profit by short weight or adulteration.

The next best test is its appearance. Good Peruvian guano is an impalpable powder, perfectly dry to the touch, of a uniform brownish yellow color, with a strong smell, like that of spirits of hartshorn, contained in ammoniacal smelling bottles. But the smell is no test; that which smells strongest may be worst, as the ammonia may be disengaged by moisture or by the addition of lime or salt.

The adulteration of guano is carried to a great extent in England, and probably will be in this country. The principal adulterations are made by the addition of loam, marl, sand, plaster, old lime, ashes, chalk, salt, moisture, and by mixture with other guano of a cheaper quality. The farmer need not depend upon the assertion, "this is a genuine article—here is the inspector's certificate." We would not give a straw for a corn basket full of certificates of analysis. The buyer must analyse for himself. Mr. Nesbit, analytical chemist, London, has just published a pamphlet from which we have condensed some very plain, short, simple rules for testing the quality of guano. As the adulterating substances are generally heavier than the guano, they may be detected by a comparison of weight and measure. To do this, get a small glass tube closed at one end, and weigh accurately an ounce of pure guano, put it in the tube and carefully mark the hight it fills—try several samples—if there is any difference, mark it. Now weigh an ounce from a sample adulterated with one fourth its bulk of any or all the preceding list of articles used for that purpose, and you will find the difference of bulk between that and the genuine, very perceptible.

Test by Burning.—Guano burnt to ashes at a red heat will leave an ash of a pearly white appearance, not varying in weight from 30 to 35 per [Pg 70] cent. of the quantity burnt. If it is adulterated with marl, sand, clay, &c., the ash will be about 60 or 65 per cent, of the weight tested, and be colored with the iron always present in the adulterating substances, and which is never found in pure guano. This test, to be accurate, must be done with a nice pair of scales and a platinum cup, which may be heated over a spirit lamp. Ten grains of the guano are placed in the platinum cup, which is held by the tongs in the flame of the spirit lamp for several minutes, until the greater part of the organic matter is burnt away. It is allowed to cool for a short time, and a few drops of a strong solution of nitrate of ammonia added, to assist in consuming the carbon in the residue. The cup is again heated, (taking care to prevent its boiling over, or losing any of the ash,) until the moisture is quite evaporated. A full red heat must then be given it, when, if the guano be pure, the ash will be pearly white, and will not exceed 3½ grains in weight. If adulterated with sand, marl, &c., the ash will always be colored, and will weigh more than 3½ grains. Even the simple burning of a few grains of guano, on a red hot shovel, will often indicate by the color whether a fraud has been committed; but we cannot particularly recommend this method, as the iron of the shovel itself will sometimes give a tinge to the ash. This might be obviated by burning the sample on a common earthen plate.

If the adulteration of guano has been made by sand, it can be detected by dissolving the ashes in muriatic acid. The sand will remain—if it is more than one per cent., it has probably been added fraudulently. As iron exists in loam, it will show in the color of the ash if that is the substance used for adulteration. If lime has been added, it can be detected by dissolving the ash in muriatic acid and separating the sand, loam and iron, if present, by filtration, and then adding oxalate of ammonia to the liquid. If it shows more than a mere trace of lime, it has been falsified.

Test by salt.—Saturate a quart of water and strain it; pour some in a saucer and sprinkle guano upon the surface. Good guano sinks immediately, leaving only a slight scum. If it has been adulterated by any light or flocculent matters, they will be seen upon the surface of the brine.

Test by Acid.—Put a teaspoonful of guano in a wine glass and add a little vinegar or dilute muriatic acid. If ground limestone or chalk have been added, the effervessence will show it. A genuine article will only show a few bubbles.

Test by Water.—The following simple plan will easily detect all the ordinary adulterations of guano. Procure a wide mouthed bottle, with solid glass stopper; fill with water and insert the stopper; let the exterior be well dried. In one pan of accurate scales, place the bottle; [Pg 71] counterpoise by shot, sand or gravel. Pour out two thirds of the water, and put in four ounces avoirdupois of guano. Agitate the bottle, add more water; let it rest a couple of minutes, and fill with water, so the froth all escapes; insert the stopper, wipe dry, and replace the bottle in the scale. Add now to the counterpoised scale, one and a half ounces avoirdupois, and a fourpenny piece; if the bottle prove the heavier, the guano is, in all probability, adulterated. Add in addition a three-penny piece, and if the bottle is still heaviest the guano is undoubtedly adulterated. By this simple experiment, a very small amount of sand, marl, &c., is detected.

If farmers will not use some of these simple tests, or employ a chemist to detect suspected adulteration; or if they will buy guano of men who have no character to lose, and who offer to sell below a price to afford them a living profit, they cannot be pitied if they are cheated.

Prepared Guano.—Never buy anything bearing that name, unless you wish to verify the adage of "the fool and his money are soon parted."

Analysis of Prepared Guano.—We give an analysis of one sample of domestic manufacture, and two British. No. 1. was offered in London and actually sold as Peruvian guano, to farmers in the south of England; just because they were so neglectful of their own interests as not to inform themselves that an article sold for $35 a ton, could not be genuine, while the regular government price remained fixed at $47. It may readily be seen by the analysis, how they were cheated into paying that price for an article of which 74 per cent. is plaster, and only half of one per cent. ammonia.

No. 1.	Gypsum,	74.05
	Phosphate of lime,	14.05
	Sand,	2.64
	Moisture and loss,	9.26
		— —
		100.00
		— —

Ammonia,	0.51

The other sample is still worse. This was sold as Saldana Bay guano, at $15 to $20 a ton. It was composed of

Sand,	48.81
Phosphate of lime,	10.21
Gypsum,	5.81
Chalk,	22.73
Moisture,	12.44
	——
	100.00
	——
Ammonia,	a trace

[Pg 72]

It would have been dear at half the price. But why? perhaps you inquire, do you give these samples of rascality in England? Just to show you what men are capable of doing there, they will probably do here—nay, have done. Here is the analysis of an article which was sold in the city of New York, under the name of *prepared guano*. The analysis was made by the lately deceased, highly respected, and eminent analytic chemist, Professor Norton, of Yale College, showing the following result.

Water,	4.35
Alumina and phosphate of lime,	7.82
Organic matter,	32.58
Insoluble matter,	26.05
Carbonate of lime,	28.76

| Magnesia, alkalies, and loss, | 0.43 |

100.00

This analysis was made by the request of the editor of the Genesee Farmer, by whom it is not only endorsed, but proof given of its utter worthlessness upon the land where it was applied. Professor Norton made the following remarks upon the subject.

"This is indeed a *prepared* article. You will observe that three tenths of the whole are water, or matter insoluble in acid, or nothing more than water and sand. More than another three tenths is organic matter; this contains scarcely a trace of ammonia or nitrogen in any form, being worth no more than common muck from a swamp. Thus we have six tenths of the guano made up of a mixture that as a gift, would not be worth carting. Nearly another three tenths is carbonate of lime, a valuable article it is true, but one which can be bought far more cheaply by the barrel, bushel or ton, than as guano. The remaining tenth contains a small quantity of phosphates, but not enough to make the mixture of much value. The parties engaged in this manufacture, should be widely exposed, for it is one of the most outrageous impositions I have ever known. Farmers should avoid everything of this nature unless it is certified to be equal to a copy of analysis shown. This stuff is not worth transporting any distance for your land. J. P. Norton."

We will now give the analysis of Peruvian, Patagonian, and Chilian guano, as determined by Dr. Anderson, chemist of the Royal Agricultural Society of Scotland, to be a fair average deduced, from a careful examination of many samples. The same results have been obtained in this country by such eminent chemists as Professor Norton, Dr. Antisell, and Dr. Higgins. We only give analysis of these three kinds, for the reason, no other of any consequence is now offered for sale in this country. [Pg 73]

ANALYSIS OF GUANO.

| Peruvian. | Chilian Fine. | Chilian Inferior. | Patagonian. |

Water,	13.73	6.06	15.09	24.86
Organic matter and ammonical salts,	53.16	54.51	12.88	18.86
Phosphates,	23.48	11.96	16.44	41.37
Lime,	— —	1.37	8.93	2.94
Sulphuric acid,	— —	— —	— —	2.21
Alkaline salts,	7.97	10.25	6.04	2.70
Sand,	1.66	15.85	40.62	7.56
	100,000	100,000	100,000	100,000
Ammonia,	17.00	18.80	2.11	2.69

It will readily be seen there is a vast difference in the value of the Chilian, and though not stated, there is as great a difference in the Patagonian, while that from Peru, owing to the fact that it never rains upon the depository, is of a uniform quality. As the principal value of guano consists of the ammonia and phosphates, it is easily calculated.

17 per cent. of ammonia is equal to 340 lbs. in

a ton of 2,000 at 12½ cents, $42.50

23.48 per cent. of phosphates is equal to 470 lbs.

in a ton at 1½ cents, 7.05

Alkaline salts, 5.00

— —

Value of a ton of Peruvian guano, $54.55

To this may be added the advantage of having these valuable substances in the best possible condition, so finely pulverized they are ready prepared for the use of plants.

It may be taken as an incontrovertible fact then, that guano is a cheap and good manure for any land and any crop which would be benefitted by the best quality of farm yard manure and ground bones. It is most beneficial on poor sandy loam, absolutely unproductive; and most profitable when applied to any land which cannot be otherwise manured on account of distance and transportation of grosser articles. The better the land is kept in tilth, the better will be the effect of an application of guano. The public may also be assured of another fact; if the guano is bought direct from the agents of the Peruvian government in this country, or of reliable merchants, who get their supplies direct from them, it will be of a uniform quality and value, as indicated by the analysis just given.

They may also rest assured, and the author of this pamphlet believes his reputation will warrant the assertion and belief, that he could not be [Pg 74] hired to puff an unworthy article, or write a book to induce American farmers, to purchase an article which would not prove highly beneficial to their best interests.

The author does know that the introduction of guano into this country is a blessing to the nation. Its general use will not only increase the wealth of individuals, but that of the body politic. Let us illustrate this point by a statement of an English writer of its advantages to that country. He says—"The importance of this question may be easily illustrated. We grow in this country about 4,000,000 acres of wheat annually. An application of two hundred weight of guano to each acre would increase the produce by six bushels, or raise the average of England from 26 to 32 bushels an acre, giving a total increase to our home produce of 3,000,000 quarters of wheat, which is of itself equivalent to a larger sum than the whole diminution of rent stated by the Chancellor of the Exchequer to have been occasioned by free trade in corn. But this is only one use to which guano would be applied, for its effects are even more valuable to green crops than to corn."

The proportionate advantage to this country would be almost inconceivably greater as our average product is far less, and the increased number of bushels per acre, far more; the produce of land as stated by Mr. Newton and others, having been raised from 3 to 15 or 20 bushels per acre.

The estimation in which it is held by some of the best farmers in the world may be judged by the increased demand in England.

The quantity of Peruvian guano annually imported has risen from 22,000 tons in 1846 to 95,000 tons in 1850, but has increased during the last year to about 200,000 tons. If the price were reduced by £2 to £3 a ton, even the present large supply would be found greatly short of the increased demand. In a single season, in 1845, when the price of Ichaboe guano ranged from 6£ to 7£ a-ton, the importation with an open trade rose to 220,000 tons. A reduction of 2£ to 3£ a ton would be followed by an extraordinarily increased consumption. Twice the present importation might be taken advantageously for the wheat crops alone. It seems to be held by the Government that the right of Peru to the Lobos Islands is unquestionable. It is, in that case, only by friendly negotiation that anything can be done. Considerations should be pressed on the present Ministry, pledged as they are to promote the landed and shipping interests. If they can persuade the Peruvian Government, by friendly negotiation, that the interests of that country as well as ours will be benefited by opening the guano trade, they will confer an important service on this country; a full supply would contribute materially to restore the prosperity of the landed interest by increasing [Pg 75] their produce at diminished cost; and it would give regular employment to about one-tenth of the whole mercantile navy of England.

Undoubtedly! an increased supply, or rather an increased consumption, would tend materially to restore, in England and in America, to build up the landed interest, by increasing the product of the land at diminished cost. If farmers could buy guano at lower prices, it is argued all would use it. Undoubtedly again! Because their profits would be greater. So great in fact, the temptation to make money out of the purchase and use of guano few could withstand "such a chance for a speculation."

But as they cannot induce the Peruvians to let them have it at a lower price, and as they can make money out of it at the present price, is it not a suicidical measure upon the part of the owners of unprofitable land, to refuse to use guano, because they cannot get it at their own price, while they can certainly profit by its use at present prices.

The Guano Monopoly.—Much prejudice has been excited against the agents and principal dealers in this country by the cry of monopoly. Are those who cry *wolf* the loudest, entirely clear themselves, of a fondness for fat mutton? The following extract from a letter of Edward Stabler of Maryland, gives a more fair, impartial view of the subject. He says; "Odious and grinding as monopolies usually become, and hard as this one seems to bear upon the agriculturist's interests, it still appears to be about as fair as ordinary mercantile transactions. The Peruvians may be considered the producers, and like our farmers and planters, may at times require advances from the commission merchant; and in proportion to the prices obtained, are his profits increased; nor does any one censure the merchant for selling at the highest price he can. Dealers, or speculators, if you please, are always censured for raising the price of guano. Is not the same thing done every day, and every hour in the day, by the purchase and sale of flour, wheat, corn, and tobacco—and is not the price of almost every article of commerce regulated in a great degree by the supply and demand? Most certainly; and so long as there is a probability of profit by the purchase and sale of this article, and just so long, and no longer, will the 'trade in second hands' continue. If the present supply is inadequate to the demand, by an almost undeviating rule in commerce, the price is enhanced, until at a point to drive the consumer from the market. This however, is not quite so soon attained with guano, under the present excitement, as with many other things. I have viewed this matter in a different light from some others, though erroneous as some may suppose, and do not think that censuring the dealers will cover the true ground of complaint, or at all tend to remove the existing difficulty. Their agency is, if I may use the term—but in no offensive sense—a kind of neces [Pg 76] sary evil; for the importer will not retail, and it suits but few of the consumers comparatively, to club together, and purchase in large quantities. The price of guano is owing mainly, if not entirely, to this monopoly in the import trade; and it would be the same thing, and a monopoly still, whether in the hands of English or American merchants; with also, about the same amount of liberality to be looked for, from one as from the other."

Is there anything so unfair in this, that we should cry out "wicked monopoly." The Peruvian government, after the revolution, finds itself deeply in debt, and greatly in want of money, and in possession of one of the most valuable fertilizing substances in the world, which the people of other governments are in want of, or rather, may profit by the use of, which she offers to sell at what she deems a fair price; and for the purpose of enabling her to borrow money for immediate necessities, as well as to pay the war debt, she has given some of her citizens—rich merchants, who can advance money, certain privileges and advantages in the guano trade, upon condition that they will send a supply to all the countries where it can be sold, and in as great quantities as they will buy at fixed prices. This is the monopoly. A parallel case can be found nearer home. The government of the United States, also incurred a revolutionary war debt, and also came in possession of an article which the people of all other countries want, and unlike that possessed by Peru, an article which they must have. Upon this necessity of life, our government has fixed a price, which any one may pay or let it alone— buy or not, just as he pleases. The government will neither sell to citizens or strangers at half price, nor let them have the use of it without pay; in fact, will not let us carry away anything of value from this property, although it might not materially injure the sale of the principal and most valuable portion, which is immovable. Such is the "guano monopoly" of one government, and such is the "land monopoly" of the other. Which is most wicked?

Of the right of each government, no honest man will dispute. That Peru has as much right to the guano upon her desert islands, as the United States has to the live oak timber in the deserts of Florida; or as England has to the codfish in the waters of Newfoundland, seems to be as clear as any right ever exercised by any power on earth. Each protect their own by hired agents, so far as they are able, to prevent dishonest men from carrying away that which each considers valuable.

If English and United States citizens have a right to go and seize upon the guano and bring it off in defiance of Peru, because the guano islands are not inhabited, then have we a right to seize all the codfish in the waters of the sea, because nobody lives there—they cannot live there—they only live on the lands adjacent, and therefore have no right [Pg 77] to anything except what they stand upon. Then by the same rule may the lands of the United States be seized upon, because they are unoccupied.

By virtue of decrees now in force, no vessel, either under the national or any foreign flag, has a right to go to the Peruvian guano deposits, without first obtaining permission from the Peruvian Government under penalty of confiscation.

Foreign vessels, furnished with government licences, are allowed to load at the Chinche Islands only.

Finally, any attempt to load vessels without the proper licences, would subject them to be seized by the Government vessels appointed to cruise off, and visit the different guano deposits, in order to prevent not only the illegal extraction of guano by foreign trading vessels, but also to prevent the natives of Peru from violating the Government orders against visiting those localities, and destroying or disturbing the birds.

Notwithstanding this cuts off the free trade in the article, it goes to show what we have always endeavored to impress upon the minds of American farmers, that the supply is inexhaustible—at least in this age and generation—and as every one grows wiser and wiser, it is probable the next will have no occasion to use such an old fashioned article as bird dung for manure. During the present, however, our advice is to every person occupying land which needs something to improve its fertility, to use guano—genuine Peruvian guano—purchased of reliable merchants—and the fewer the better between the importer and consumer.

The Quantity inexhaustible.—By those surveys, the quantity was ascertained to be upwards of TWENTY MILLIONS OF TONS. As this must appear so enormous as to be almost incredible, we present the annexed cut, supposed to represent a vertical section of one of the Chincha islands and the depth of the deposit according to the government surveys. The paralel lines at the bottom represent the level of the water—the crooked line above, the surface of the rock; its position having been ascertained by boring and observations of the surveyors. The rounded line is the surface of the island as it now appears; all between that and the rock being guano. The almost perpendicular line at the left hand, 100 feet high, is the rock at which ships lay to take in cargo. The space under the dotted line show a comparison of the quantity taken away, as it relates to the whole upon the island. The well hole represented in that section was dug some fifty feet deep to prove the guano was of equal quality at the bottom.

The Chincha Islands are three in number; not remote from each other or differing very materially in size or general feature. The Geological [Pg 78] formation presents the appearance of masses of rock jutting out above the surface of the ocean — and occasionally rising nearly perpendicularly to a height of from 50 to 100 feet. At a distance, the islands present to the eye a somewhat conical form; owing probably to the greater deposits of guano in the centre; and all appear equally rich in quantity and quality.

The "North Island" is estimated to be about 300 feet at its greatest elevation; it is about 1½ miles in length, and from 1/2 to 3/4 of a mile average width. In sailing round them, the guano appears to many places to extend to the water's edge.

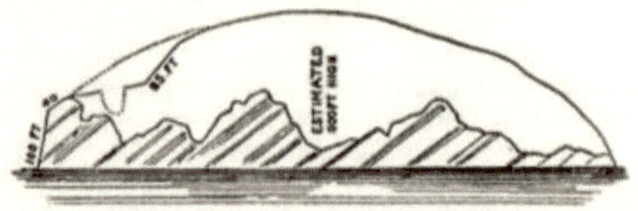

SECTIONAL VIEW OF THE NORTH CHINCHA ISLAND.

All the guano islands are uninhabited, except by the laborers, mostly Indians or poor Chinamen, who are employed in the work of digging, carrying and loading the guano into the ships. When a vessel is ready to take in cargo, she is moored alongside of the rocks almost mast head high, from the top of which the guano is sent down through a canvass shute directly into the hold of the ship. Thus several hundred tons can be put on board in a day. The trimming of the cargo is a very unpleasant part of the labor. The dust and odor is almost overpowering; so the men are obliged to come often on deck for fresh air. The rule is to remain below as long as a candle will burn; when that goes out, the air is considered unfit for respiration. If the labor had to be performed by a Yankee, he would think it unfit at first; and thereupon set his ready wit at work to construct a machine to spread the guano as it fell, from one end of the hold to the other. The guano in position upon the island, is so compact it has to be dug up with picks. It is then carried to a contrivance made of cane, at the edge of the rock, which conveys it into the canvass conductors. The mass is cut down in steps, receding and rising from the point of commencement, and has not yet attained a depth of 100 feet, and with all the labor of hundreds of men digging, and numerous ships carrying away to the several countries using it, there is but a bare beginning of removal made upon the mass upon one island only, as may be seen by reference to the diagram. [Pg 79]

Supposing like many others, the supply of Peruvian guano was like the Ichaboe, destined to run out—that is all be dug up and carried away; we inquired of an intelligent captain of a ship just returned with a load, how long it would be before the supply would be exhausted. "Exhausted!" said he, with a look over the gangway, as much as to say how long would it take to exhaust the ocean with a pint cup; "why not in one hundred years, if every ship afloat should go into the trade, and load and unload as fast as it would be possible to perform the labor; no, not from the Chincha islands alone. Exhausted! they never will be exhausted." With due allowance for the captain's enthusiasm, we may be very certain from the government surveys, the quantity is so great, that no probability exists of the supply being exhausted until all the present inhabitants of this earth have ceased to move upon its surface. We may be certain of another fact; that unless we commit a great national wrong upon Peru, by seizing upon some of her guano territory; a thing which the sober second thought of this nation will never sanction; we shall not be able to obtain the article only through her government agents, at such prices as her rulers think proper to affix to it. While the demand and the result of the use of guano continues as at present, there is not much probability of any material change.

The Peruvian Government are, of course, anxious to sell all that the world want, and are willing to pay for at remunerating prices. The Peruvian minister, in reply to the Secretary of State at Washington says:—"The Peruvian Government, in leasing out its rights and interests, as a proprietor of the article, adopted the only system that was supposed likely to create a demand for guano; while, on the other side, it was bound to leave the consignment as security, in the hands of those persons who had hazarded their capital in meeting the heavy expenses attending the process of freighting, and in making the advances which were required to facilitate the exportation and construct the depots. Far from establishing a selfish monopoly, which would have proved injurious to its own interests, or fix a high, deliberate, and conventional price upon the article, it has only aimed to secure a net profit, reduced to the lowest possible standard, exceeding very little the actual amount of expenses; and there have been accounts of sales rendered exhibiting both loss and damage.

"The guano, therefore, is not monopolized; the government as the proprietor, has forwarded it, on its own responsibility, to those markets where it was in demand; selecting as consignees, as it was natural and proper it should do, those persons or houses who have advanced the capital necessary to defray the expenses; and, as these are much greater in all cases of remittances to England, and it follows that the sale of the article in this country is at the rate of ten pounds sterling [Pg 80] per ton, the net profit has been less than what is realized in the United States, where the farmers obtain it at lesser prices. Nor has my government imposed any restrictions, duties, or determinate value on the exportation of guano, although it might and could do so with perfect propriety; because such action would have militated to the detriment of its own interests as the proprietor of the article. Its object has been to send it to those markets where it was in demand; because, as it had not yet become an object of decided and positive interest to the consuming world, and there being no certainty of its attaining saleable prices, to create a market as it was impossible for individuals to send to Peru for supplies, with any prospect of even moderate profit."

This is a fair statement of the case; and ought to be perfectly satisfactory to the consumers. The disposition of some men to create prejudice against the government of Peru, or the agents who sell guano in this country, because the price is too high, is a wicked one. Men can make money by purchasing at the present prices; and the owners of the article think they cannot make it by selling at a lower price. We have heard it urged as a reason why it should be sold at lower prices, that the agents and merchants engaged in its sale are making fortunes. So are flour merchants—so are farmers who grow the wheat—but that is no reason why it should be sold lower.

With all our heart, we wish the Peruvians would give us guano at half price; but because they will not, there is no reason why the people of this country should refuse to use an article which will most assuredly make them grow rich faster than those who are engaged in selling it.

WHAT IS GUANO? — ITS HISTORY AND LOCALITY. — AMOUNT AND VALUE.

Guano is the concentrated essence of fish-eating birds excrements. It, is found in the condition of a dry powder, of a brownish yellow color, not unlike in appearance to Scotch snuff; with a pungent strong smell of ammonia, distinguishing it from any other substance. It is found in various parts of the world, upon desert headlands and islands of the Atlantic and Pacific Oceans, where the birds have had undisputed possession for countless ages of time. The island of Ichaboe, on the Coast of Africa, furnished a good many cargoes, a few years since, most of which were taken to England; a small supply was imported into the United States, and sold and known as African guano. The quality was fair The deposit upon that island is quite exhausted — in fact it was all carried away within a few months after it became generally known — some of the last cargoes being of little more value than rich earth. It is said that a new deposit, which is nothing more than dry bird dung, has already been gathered and taken to England. No doubt cargoes of similar ma [Pg 81] nure might gathered from the Florida keys; and although it would be a valuable manure, it is not guano — that is formed by the chemical action of a dry atmosphere, during time's long ages.

Anagamos Guano. — This is also of a character similar to "new Ichaboe." It is rich in ammonia, but contains no lime or sulphuric acid, and less phosphates and alkaline substances than Peruvian, and more sand. The supply of this must be very limited, as it is a recent deposit and has to be gathered by hand from the rocks.

Bolivian Guano. — This as its name indicates, is from the coast of Bolivia, on the west side of South America. It was thought at one time to be fully equal in value to Peruvian, but some subsequent importations of almost worthless cargoes, have proved the deposit to be very variable in quality, or else purposely adulterated, which has had the effect to destroy confidence in all bearing that name. The belief of the writer is, that it was not adulterated, but owing to the fact that it is found in a latitude where it does sometimes rain, or where it is liable to be drenched by sea spray, that portions of it are injured in that way; so that a ship may have one portion of her cargo of the best kind, while the remainder is hardly worth the freight. The deposit is not large.

Chilian Guano.—The most of that imported into this country under this name, has been of a very inferior quality, and having been recommended by those interested in its sale, as having come from the same coast as that of Peru, and of equal value, and proving almost worthless, has deterred many from making another trial. Although there is a small supply of Chilian Guano, which is gathered from the rocks in pale yellow masses, some of which has been sent to England and this country, which is equal to any ever discovered in any part of the world, yet the great bulk of the deposit is so inferior that Chilian guano will never meet with universal favor. In fact, some of the stuff which has been sold under that name, is unworthy to be called guano.

Patagonian Guano.—Of this kind, larger quantities have been imported than any other beside Peruvian; and it has generally been sold at higher prices than its value as a fertilizer would warrant. Owing to the fact of its being deposited in a latitude of sunshine and showers, both of the utmost intensity; it never comprises the valuable qualities always found in that where rain never was known to fall. Besides the deterioration of the elements, samples of some cargoes of this guano have been found to contain upwards of 30 per cent of sand—in one case 38 per cent. It is said, however, that some of the deposits contain considerable quantities of crystalized salts of ammonia, magnesian phosphates, rich in ammonia, but which have been rejected by masters of vessels taking in cargoes, under the supposition of its being sea salt and calculated to injure the sale and value of the guano. It is believed that there is a [Pg 82] larger supply of this than any other guano, except Peruvian, but as no certain reliance can be placed upon its quality or value, it never will be extensively imported into the United States.

Saldana Bay Guano.—Considerable quantities of guano under this name have been taken to England, and upon land and crops requiring phosphates more than ammonia, has been pronounced a superior article. But the fact is, it is found in a climate similar to the Patagonian, and, consequently, like that, must have a great portion of its ammonia washed out, leaving almost its only value as fertilizer, in its phosphates; which undoubtedly exist in large proportions, but not as cheap as may be procured from other sources. The foregoing comprises all the kinds of guano known in commerce, except the Peruvian, to which we shall devote an entire chapter.

PERUVIAN GUANO—ITS LOCATION—OWNERSHIP—QUANTITY—VALUE—HOW PROCURED.

This is not only the most valuable, but is found in the largest quantities of any other guano known. That which has been sent to this country and England, in such quantities within the last ten years, was taken from the Chincha Islands, which are situated between latitude 13° and 14°, and at about twelve miles from the coast of Peru, in the bay of Pisco. The great value of the Peruvian guano, arises from the fact, *that rain never falls upon the islands where guano is found*. The air is always dry, and the sun shines with intense power, sufficient to evaporate all the juices from flesh, so that meat can be preserved sweet without salt. The waters surrounding these islands may be said to be literally alive, so full are they of fish. Almost as numerous as the fish, are the birds which satisfy their voracious appetites upon this finny multitude, until they can gorge no more, when they retire to the islands to deposit their excrement, composed of the oily flesh and bones of their only food, until the mass which has been accumulating for thousands of years, is so great as almost to exceed human belief.

Humbolt, in his history of South America, states, some of these deposits are 50 or 60 feet thick. Many have thought this the "romance of history," but the actual surveys made by the Peruvian government five or six years ago, have proved that the guano in many places is more than twice that depth; and as there is good reason to believe, and as may be seen by the diagram on page 79, it is probably 300 feet thick in some of the depressions of the natural surface. And this has been accumulated by an annual aggregation, so slow as to be scarcely visible from year to year, until the quantity now exceeds 20,000,000 of tons.

As before stated, the Chincha islands are three in number; the Lobos islands two; these are situated off the north part of the coast of Peru. [Pg 83]

If the right of Peru to the guano is to be disputed, let it be done by national vessels and not by armed privateers. If farmers are convinced that we have made true statements of the value of guano in renovating the poor and worn out fields of America, let them purchase at once. The only question to ask is not whether we can go to the Lobos Islands to get guano—nor whether it would be better to buy it of government agents, or speculators on private account, but

DOES GUANO PAY?

Because, if it does pay, that is, if the farmer can buy guano at present prices, and realise an increase of crops more than enough to pay the expense, it does pay. We think we have shown this fact by incontrovertible evidence. If the first crop pays for the guano and no more, the farmer has a certain profit in the improved condition of the land. If the first crop does not pay, the land will be enough better to pay cost. Upon this point, Mr. Mechi, of England, whose name has become world wide known as an improver of the soil, says; "Whether guano will pay, depends upon the condition of the soil. On poor exhausted soil it is a ready and cheap mode of restoring fertility. I used it extensively when I first began farming, and when applied to the grain crops at the rate of two to three cwt. per acre, it paid well; but now it has lost favor with my bailiff, which is easily accounted for; my land being at present so well filled with manure, nitrogen or ammonia, that we can grow ample crops without it. When the land only yielded two to two and a half quarters of wheat per acre, it was grateful for guano; but now, with a produce of five quarters, there is no necessity for its use. Or rather, the increased supply of farm manure supplies that necessity."

This is exactly what we have aimed to impress upon our readers; that it will pay in the crop to which it is applied—it will more than pay in the soil, because it will bring it into a condition of permanent fertility. It will pay best upon the poorest soil; because that which was absolutely barren, becomes fruitful as soon as dressed with guano. It will always pay whenever and wherever applied to any soil in a fit condition to be benefitted by manure. It will make not only the soil rich, but whoever uses it to any considerable extent. It will pay best when used in the condition in which you buy it, with no additional labor or expense except breaking the lumps. If it is sown broadcast, not to exceed 400 lbs. per acre, and plowed in so deep it will not be disturbed by any subsequent cultivation of the crop to which it is applied, it will most certainly pay in that crop or the succeeding one. It will pay upon all plants to which it has ever been applied. Notwithstanding it will pay best *in* the soil, it will pay well *on* it as a top dressing, [Pg 84] if combined with absorbents of ammonia as directed in these pages. [2] That it has paid in ninety nine cases out of every hundred where it has been used, the author is well convinced, and equally well convinced that many may profit by reading what he has here said upon the subject, and with that feeling, these pages are commended to all the cultivators of American soil.

[2] Upon this point, see Mr. Burgwyn' letter in the appendix.

[Pg 87]

[Pg 86]

[Pg 85]

APPENDIX.

SUCCESSFUL EXPERIMENTS WITH GUANO ON LONG ISLAND.

Since the body of this work was in type, the following letters have been placed in our hands. They contain so much valuable information we are induced to append them. It will be seen by the dates, that they give the results of the most recent experiments. The names of the writers will be recognized as those of reliable, practical men.

> **Letter from Seth Chapman Esq., of Jamaica. — 700 lbs. of guano to the acre, profitable — Lasting benefits of one application — Advantage of top dressing grass lands with guano — Benefit of guano to all Long Island soil — Great benefit on turnips.**

"Jamaica, L. I., Sept. 13, 1852.

Mr. Theo. Riley, Esq., Dear Sir: — In reply to your inquiry relative to the use of Peruvian guano on Long Island, I would say, forming my opinion from experience and observation that the mode of tillage — the rotation of crops, and the way of applying guano — are about as follows: Commence with corn, which is usually on green sward, after being mowed and pastured from four to six years. First, plow in the spring as soon as the frost is out of the ground, which is generally about the 20th of March. Prepare the ground for planting the 1st of May, by harrowing well two or three times. Before the last time harrowing, apply about 250 or 300 lbs. of guano to the acre, sown broadcast, and then mark out with plow, or lace, about four and a half feet apart, each way; apply a small quantity to the hill, one third of a gill is as much as will be safe, and that should be in the form of a ring about a foot in diameter, and the corn dropped in the center, otherwise it will be likely to kill the corn by the sprouts coming in contact with the guano when they first start. It will not do to put the guano in the hill and plant the corn upon it. It was not uncommon for farmers to have to plant their corn all over before they become acquainted with its effects; but as using it in the hill, in a pure state, is generally attended with some risk, it is the practice in this vicinity to use yard manure, at the rate of one third or half a shovelful to the hill; but as that manure is generally weak, they have adopted the very excellent plan of sprinkling say 50 lbs. of guano to a wagon load (30 bushels) of manure. As we cart the manure in the fall to the field where it is intended to be used the following spring, (1) the guano can be mixed through it with but [Pg 88] little trouble, when it is turned and broken up just before use. It adds very much to the value of the manure, as the difference of harvesting plainly shows. Muck or pond dirt could be used in the same way, in place of manure. Some apply it about the hill at the time of hoeing. It should not be thrown on top, but sprinkled around the corn at the rate of half a gill per hill. After corn, we sow oats, or barley, or plants potatoes; if oats, plow once, sow 150 or 200 lbs. of guano, and two bushels of oats to the acre, and harrow in together. It pays well to use guano for oats, as the crop of oats will be doubled on ordinary lands; 50 and 60 bushels is frequently obtained, and the difference in the straw, is worth the expense of the guano. (2) Barley is not much sown; it would require a little more guano, say 50 lbs. additional. Potatoes, (Mercers) we plant from middle of March to first of May, after sowing broadcast from 400 to 600 lbs. of guano per acre, plowed in and harrowed over; then mark out with plow three feet apart, drop in drills about a foot apart. Some prefer it in the drills, at the rate of what they can grasp in one hand to a pace of two and a half feet; it should be sprinkled so too much will not come in contact with the seed. After oats or potatoes, sow wheat, about first of October; if on oats, plow as soon as the oats are off; when ready to sow, apply from 500 to 700 lbs. of guano per acre, cross plow, and your ground is ready for the seed. As to the varieties of wheat, there are several kinds used; the Mediterranean is the most popular at present—one and a half bushels is generally sown to the acre, and the land laid down to grass, with timothy and clover. Some apply less at time of sowing, and add 100 or 150 lbs. per acre in the spring, just as the grass is starting, say first of April. If wheat is sown after potatoes, about the same treatment is given, except 100 lbs. less guano will answer. Some harrow in guano, instead of plowing it under; but experience shows that it is much the best to plow it in, as the virtue remains in the ground much longer, by being covered deep. Peruvian guano will produce the best wheat of anything we can use, even if we should go to double the expense with other manures. Crops of 30 and sometimes 40 bushels have been obtained to the acre with guano. The average crop of wheat on the Island, is not over 18 bushels per acre, and with 700 lbs. of guano plowed in pretty deep, the land can be mowed about as long as from an application of stable manure. But as hay is a most important crop, after it has been mowed for two or three years, it is considered profitable to top dress with about 150 lbs. per acre; this will increase the crop from one ton to two per acre, if a fair season, and can be mowed two or three years longer. Rye is sown in many instances, in place of wheat; it gets the same treatment, except half the quantity of guano is only used. Buckwheat requires about 100 lbs. of guano to the acre, more or less, according to the state of the land. [Pg 89]

For ruta baga turnips, there should be 600 lbs. sown to the acre; plow twice and harrow well after sown. After you have hoed them out, give them a light top dressing of more guano. I have raised at the rate of 700 bushels, managed in that way, to the acre. We have had one of the most extreme drouths the present season I ever remember. Crops on which guano was used, have suffered less, and are now yielding better than where stable manure has been used. This is quite different from the opinion that some have formed, as to guano requiring a wet season. To prepare guano for use, it should first be sifted, to separate the lumps, so that they may be pulverized, then dampen by sprinkling with water, and mixed through with a shovel. This should be done a few days before you wish to use it, so as to allow the dampness to strike through uniform. (3) I have not had any experience with compost, or using it on garden vegetables, or plants, except I know it should be used in homeopathic doses, or it will destroy more than it will produce. As to the soil, guano answers well anywhere on Long Island, although some parts of the Island has a very different soil from others, with one exception; that is, it is all hungry for manure. I therefore do not know the kind of soil it is most applicable to, since it seems to suit all kinds.

Seth Chapman."

Note 1. This practice of hauling manure to the field in the fall, is the worst of all the foolish old fashions of farmers. To preserve the virtue, of manure, it requires housing about as much as hay. In fact, it is doubtful which would lose virtue fastest, a pile of hay or a pile of manure, exposed to the storms of winter. It is no wonder that it becomes necessary to mix guano with it, to replace that which has evaporated during its long exposure to sun and storm.

Note 2. This increase of straw, is seldom taken into account in speaking of the advantage of an application of guano; yet, as Mr. Chapman says, it is worth enough in the vicinity of a market, to pay the whole expense. It is also valuable in the interior for forage and manure.

Note 3. This is an error. Guano should not be damped unless with water saturated with salt, copperas, or a liberal sprinkle of plaster over the pile.

**Letter from Seth Ravnor, of Manorville to Mr. Chapman.
Successful experiments on grass, oats, corn, wheat and rye.**

"Manorville, Sept. **8, 1852.**

S. Chapman, Esq.—Dear Sir;—I have received your circular proposing to gather information from practical farmers of the results from the use of guano, and to have the same published for general circulation. Conceiving the object to be a very laudable one, I will give the result of a few experiments tried with Peruvian guano by myself, and others which have come under my observation; but in doing so I think [Pg 90] it would be of great utility to state what kind of soil the guano was applied to. Not being a professor of geology, I can only use such terms as are familiar with farmers generally. The soils in this vicinity are heavy loam, sandy loam, sandy, and occasionally some heavy clayey soils.

First, as to the nature of guano. It is generally considered to be more of a stimulant than an enricher of the soil, if applied in its natural state, and much more durable to be plowed in than to be harrowed in; and as far as I have tried it, I have not found it to be injurious to soils—or as some call it, 'kill the soil.' In the year '49 I applied on the first of April, 176 lbs. per acre on sandy loam grass ground—yield, about half a ton more than the acre adjoining. Same year applied about 150 lbs. to the acre, on four acres of oats, same kind of soil, and the estimated increase was 20 bushels to the acre. In 1850 plowed under 400 pounds per acre, for corn, estimated increase, 15 bushels of ears. The season was rather unfavorable for corn. In '51 composted six bushels charcoal dust to 100 lbs. guano, and plowed under for wheat, at the rate of 500 lbs. of guano so composted, to the acre, and top dressed with 100 bushels of leached ashes—yield, 20 bushels. One of my neighbors applied for three years in succession, 100 lbs. harrowed in with rye, on two acres light sand—yield, 14 bushels to the acre; 10 bushels more than the acre adjoining. On the fourth year he sowed the same ground without guano—- yield, 4 bushels to the acre. We see by this, that the crop used the whole strength of the guano. Another neighbor applied one ton to two acres, heavy loam; plowed under and sowed with turnips (common Russian)—yield, 1,300 bushels—estimated increase from the guano, 600 bushels. People in this section of the Island are agreed in this—plow under guano for durability, and harrow in for present benefit, or present crop. For wheat, 500 lbs. plowed in is considered a full dressing per acre. The same for corn. For oats, 200 lbs. harrowed in. For buckwheat, 100 lbs., and 200 for barley. One tablespoonful applied in a hill, for corn, is quite enough, and that requires to be put some six inches from the seed; otherwise it will kill it. Some have lost acres by putting their corn on that little quantity; the only safe way to apply in the hill for potatoes, is the same as for corn. I have come to the conclusion from what experience I have had with the article, that it answers the best purpose to use it for spring crops, in the manner above stated, or compost it with charcoal dust, or well decomposed pond mud, to absorb and retain the ammonia, it being very volatile in its nature. I have not written this for publication; I have only thrown out a few hints for you to embody.

Seth Raynor."

Although the above was not written for publication, we prefer to give it just as it was written, in the plain style of one farmer to another. [Pg 91]

> **Interesting Letter from Edward H. Seaman, Esq., Sec. of Queens Co. Ag. Soc.—Successful experiments since 1847—Great increase of straw and wheat—Harrowing in guano, 300 lbs. to the acre, produced 41 bushels of wheat. Increase, seven bushels for each 100 lbs—Thirty bushels of wheat per acre on an old worn out buckwheat field—Advantage of guano in drouth—astonishing effects from top dressing grass.**

Cherrywood, Sept. 11th, 1852.

Mr. Seth Chapman—Dear Sir,—I forward according to request, the results of several years use of Peruvian Guano, upon my farm at Jerusalem, Long Island.

The first decisive benefit from guano that I shall notice, was obtained from using it for wheat, as a top-dressing. In 1847, October 1st, I took a field containing 6 acres of oat stubble, on which I used some manure, all over the field; top-dressed with Peruvian guano, at the rate of 300 lbs. per acre, sown (fortunately just before a storm,) upon the furrow and harrowed in with the wheat. Four acres of the field were sown with the old-fashioned red flint wheat, which requires more manure than any other kind among us. The rest of the field was sown with a soft white hulled wheat, the name of which I do not remember. July 5, 1848.—Harvested said field—Red wheat yielded well from straw, 14 sheaves to the bushel—white wheat 20 sheaves to the bushel—straw very large and thick. Had 164 bushels of wheat, or 41 bushels per acre; and 58 bushels of white wheat or 29 bushels per acre; without the guano I think I could not have obtained much over 20 bushels per acre.—1848, Oct. 2. Again sowed wheat upon a six acre lot of oat stubble; seed red flint wheat—manured about the same as previous year—used 300 lbs. guano per acre, as top-dressing for 4 acres and moss bunker fish dirt at the rate of 10,000 per acre upon the two acres, sowed upon the furrow, and harrowed in just previous to a storm—Harvested the

10th of July 1849. The straw very large, and wheat heads long, but grain very much injured by fly or weevil—very little difference between fish and guano top-dressing; yield 188 shocks—175 bushels; not quite 30 bushels per acre. Same ground would not have produced more than 18 to 20 bushels wheat per acre without the guano—or some other more expensive manure. 1849. Oct. 3. Sowed wheat upon oat stubble field; soil thin and gravelly upon part of the field—used some barnyard manure, but not as much as previous year. Top-dressed with 300 lbs. guano and 12 bushels ground bones per acre—Harvested 12th July 1850—Yield of 5½ acres 160 shocks; injured some by weevil, and shrunken, but had 145 bushels or twenty-six bushels per acre. This ground would not have yielded fifteen bushels per acre without the guano. But the most decisive result was obtained the next year, upon an oat stubble field of six acres, a part of which had been cropped, for perhaps 15 years, nearly [Pg 92] alternately with rye and buckwheat; (sometimes a crop of each in one year.) The whole field seemed so far exhausted that we had failed to get a crop of corn or oats from it after two different trials; and I underwent no small share of ridicule from my neighbors, while preparing it for wheat. Remarks like the following were of daily occurrence—"Ah! Seaman you will fail this time." "You have not got your old highly manured fields to exhaust this time by your stimulating stuff!" "We shall now see whether guano is good for anything—this will be a fair test, because the land will not produce anything without it, &c." "You may get about 12 bushels of wheat per acre; we shall see." All agreed however, that if wheat did grow, guano should have the credit for it.

Well, we prepared the ground in about the usual manner, except perhaps plowing a little deeper than in former years. A small quantity of manure was plowed under, and a top dressing of ground bones given and sowed about the last of September—2 acres with Mediterranean and 4 acres with the red flint wheat—but owing to a scarcity of the article, could only get about 420 lbs. of guano, which was sown across the field upon not quite 3 acres, covering some of each kind of wheat; it was sown upon the furrow, and harrowed in with the wheat as usual. In 1851, April 11th, top dressed the whole field with guano, at about 200 lbs. per acre; harvested about the 8th July. The 2 acres of Mediterranean yielded 61 bushels; flint wheat straw very large, and thick upon the ground, but grain much injured by the weevil; yielding an average of 23 bushels per acre. I may remark, that where the guano was applied in the autumn, the crop was quite one third greater than where it only received the spring dressing. The last year I managed much in the same way, except that I fell short of manure, and depended entirely upon guano and bone upon a part of the field, from which part, though I have not yet threshed it, I think I shall get 18 to 20 bushels. The rest of the field was very large and considered the best between this place and Brooklyn, on a road of 25 miles in length.

My *good luck* (1) at wheat growing is now a conceded point. Now for other crops—for corn I have not been very successful; generally mixing some guano with earth in the hill at the time of planting and getting but few plants to stand; these, however, generally have been heavily eared. By mixing previously with charcoal dust I think this burning of the seed might be avoided. (2)

For buckwheat, I used 120 to 150 lbs. per acre, sown upon the furrow and harrowed in with the grain. For barley, 150 to 200 lbs. per acre; oats 100 to 120 lbs; turnips, 600 to 700 lbs. plowed under a short depth, previously to forming the drill; and I find a decided profit in using guano for all the above crops. I have seen a field of corn the present season very greatly improved in earing by the application of about 150 [Pg 93] lbs. of guano, mixed with 5 parts charcoal dust, and thrown around the hills a few weeks since during a rain storm.

I have also used guano and charcoal dust, five parts coal to one of guano, in my garden, the past season, and found the beds thus dressed stood the extreme drought better than other parts of the garden. One more case of my own and I am done. In 1851, I sowed about 90 lbs. of guano, on a piece of meadow or mowing ground, covering a little more than half an acre, from which the timothy and clover was nearly gone; I took 3 lands across the lot, leaving about 20 feet between each land. Where the guano was sown, the timothy grew large and thick and bore the drought, and yielded about one and a half tons per acre; while the rest of the field did not produce more than half that amount, and that of an inferior quality of grass. The corn upon the same field the present season, shows plainly a better yield from the above top-dressing. From observation and experience, I would recommend the mixing of guano with charcoal dust, equal parts, or five parts coal to one guano, It is much more pleasant to handle when thus mixed, being completely deodorized and rendered much more enduring as a manure, by retaining the ammonia for several years, instead of allowing the greater part to pass off the first season, as is the case when applied in a crude state, especially as a top dressing.

Prepared or decomposed muck if used with guano as a retainer of the volatile gases, in all cases where it can be conveniently obtained especially in soils where evaporation is so rapid as it is in most parts of Long Island, will pay.

That like produces like, is a favorite maxim with me—that it is necessary to replace the matter, both organic and inorganic, which we take from the soil in the form of crops, of various kinds—that by supplying the necessary chemical ingredients, we shall be able to draw a great proportion of our crops from atmospheric agents—that the necessity for using such an immense amount of organic matter as we now use in the shape of barn yard and stable manure will be partially overcome—that a great saving of expense will thereby ensue—that guano is one of the most active agents to effect such a result I am fully satisfied, not sufficient perhaps of itself, but highly useful even in a crude state—and capable when skillfully combined with others, to effect an entire revolution in our system of agriculture.

If you think the above worth an insertion in the pamphlet you spoke of, you are at liberty to insert it—if not, you will please return the letter to me, as soon as convenient, and if you think it will pass off any better, you may affix the following signature to the communication.

Edward H. Seaman, Recording Secretary,
Queen's Co. Agricultural Society.

[Pg 94]

Note 1. —Yes, that is the word—*good luck*—it is all good luck. It is astonishing how many farmers there are in this country who will stand with their hands in their breeches pockets, fumbling idle dollars, while a neighbor expends his for guano, and produces a fine crop of wheat upon an old worn out buckwheat field; at which they stare in stupid wonder at the good luck of the thing.

What a pity it had not been the good luck of such men to have been born with common sense enough to profit themselves by their neighbors good luck.

Note 2. —It would be far better to sow the guano broadcast and plow it in, or scatter it in drills and turn a light furrow on it before planting.

"Hempstead, Aug. 27, 1852.

Seth Chapman, Esq. —Dear Sir: —I believe I was the first person in Queens County using guano; having imported some from England in the ship Yorkshire, in 1842. This was from the Ichaboe Islands. I have since used nearly all the varieties, and consider the Peruvian the cheapest and best.

In applying guano, I think by making a compost, the greatest benefit is derived; say one peck of plaster, one bushel of loam, two of saw dust, mixed up a month or six weeks before using. From 100 to 200 lbs. of guano is enough for a crop of oats or buckwheat. I have not found it to succeed with corn or potatoes; probably from being accompanied by a dry season. The best wheat I ever raised was from using 350 lbs. to the acre, composted. This was on a light soil, and returned 31 bushels to the acre, on seven acres, weighing 62 lbs. The grass was poor after it. As a top dresser, I have used 200 lbs. per acre, very early in the spring, on half a lot, which mowed more than half as much again as the part not dressed. One of my neighbors has used 300 lbs. per acre, plowed in for potatoes; the yield, good, so far, having just commenced digging.

John Harold."

We might give much more evidence of the same kind, to prove that every barren acre upon Long Island, might be made productive by a judicious and profitable application of guano; but if there are any persons, who, after reading these pages, are still doubting, we must say they are most incorrigably determined not to profit by the experience of others. To such it would be useless to say more.

> *Successful Experiment with Guano as a Top Dressing on Wheat, in North Carolina.* — On Page 17, we gave some account of the application of guano by Henry K. Burgwyn, Esq., since which, we have been favored with the following letter from his brother, T. Pollock Burgwyn, written, as will be seen, not for publication, but simply to give the party from whom he purchased the guano, a detail of his success. [Pg 95]

"*New York, Sept.* 20, 1852.

Messrs. A. B. Allen &. Co.—Dear Sir:—Having promised that I would furnish you with the result of my application of the 21 tons of guano which I purchased of you last winter, I proceed now to do so, and give you full liberty to quote my experience in favor of the use of that most invaluable manure, to all who are anxious to profit by the experience of others without incurring any risk of their own. My object, and it should be that of every one who has used guano, is to extend the knowledge of its great value to any owner of poor soil, like the worn out plantations of North Carolina. I applied 20 tons of this guano as a top dressing to a field of 200 acres, which had been seeded in wheat under most unfavorable circumstances. At the time of application, so unpromising was the appearance of the growing wheat, that my manager and myself thought it almost a waste of money and labor to try this experiment, (1) but as the rest of my crop did not require any manure, I resolved to see what would be the effect. I am confident the field would not have averaged, without the top dressing, seven bushels per acre—it yielded rather over 13 bushels, besides securing to me a full setting of clover. (2)

My mode of application was as follows; to each 200 lbs of guano I added two bushels of ashes and a bushel of plaster mixed intimately, and then sown broadcast, at the rate of six and a half bushels per acre, harrowed in with a light harrow. This application was made in March, and the early part of April, and in less than three weeks after the application, the wheat had undergone an entire change, from a yellow, sickly color, to a dark luxuriant green. The application had evidently infused new life and vigor into the plants, and as the result proved, very nearly or quite doubled its product. So much for the crop of wheat; but what was still more valuable to me, in my system of farming, it likewise secured for me a full crop of clover, which would certainly have failed but for this application. I also applied one ton of this guano mixed in the same way, to a small field of oats. I plowed this under with a small plow, together with the oats; the result was equally gratifying. My chief object in this last experiment, was to secure me a small field of clover, near my stables, and in this I fully succeeded; which I feel assured I should not have done but for the guano. My brother and myself have made various experiments of late years, with guano, and concur in the testimony of all those who have tested its value, carefully and judiciously, in pronouncing it to be the most expeditious renovater of the soil within the farmer's reach; and exclusive of the farm yard, the most economical of all manures. In proof of my conviction of its value to me, I shall this fall give you an order for 20 or 30 tons more. I will [Pg 96] only add that I consider every wheat grower who would study his own interest, will find it by trying similar experiments.

T. Pollock Burgwyn."

Note 1. In a subsequent conversation with Mr. Burgwyn, he stated a fact which makes this point much stronger. After ordering the guano, he left home, giving his farm manager orders to apply if to that particular piece of wheat as soon as it arrived. Owing to the fact that the seed was injured—that the land was in a very unfit condition from poverty and drouth to produce a crop of wheat, it had assumed such a miserable appearance before the arrival of the guano, that the manager wrote to Mr. B. his opinion of the utter folly of applying anything so expensive to a crop already struck with death. Not imagining how very unpromising was the prospect of success, Mr. B. immediately wrote to him to go ahead as directed. Before the application was completed he returned home, and his first impression was to stop the work at once and give up the field as lost; but on examining the effect upon that part where the guano was first applied, he found it had already infused new vigor into the plants, for they had put off their sickly yellow color, and taken on a vigorous green; and therefore he decided at once to go on, which as will be seen by the result, was a most valuable decision.

Note 2. From personal knowledge of this very field, we are confident it would not have yielded without the guano, one half of seven bushels. It is a flat surface, clayey loam, and badly affected by winter rains, and such freezing and thawing as it had during the last severe winter. Besides it was a few years since, when it came into the possession of Mr. Burgwyn, one of those old worn out, skinned-to-death places, so common in that State, which all the deep plowing and good farming of that gentleman had not been able to restore, until he luckily hit upon guano; which notwithstanding the most unfavorable circumstances, has given him conclusive proof of its inestimable value. To say nothing of the ten bushels of wheat per acre, which we are confident he gained, the clover is worth more than the guano cost; and without it, one might almost as soon expect to grow clover upon Coney Island beach, as upon that field.

This letter contains testimony of inestimable value. It comes from a gentleman of intelligence and careful observation, who is devoted to his profession of a farmer, and who has been one of the most successful renovators of worn out plantations in the south, and it comes very opportunely to give our work an appropriate Finale.

www.ingramcontent.com/pod-product-compliance
Lightning Source LLC
Chambersburg PA
CBHW031420210526
45464CB00005B/1980